Lightning in His Hand

The Life Story of Nikola Tesla

by

Inez Hunt & Wanetta W. Draper

About All Pikes Peak Reads

Through All Pikes Peak Reads (APPR), the residents of El Paso County, Colorado, engage in historical, cultural, and literary exploration of selected works that present significant issues pertinent to our society. The use of scholars and educational presentations is key to the success of the project and to the continued participation of educational institutions serving students of all ages. More than 50,000 people participate in the program each year.

APPR brings programming to El Paso County, in its entirety, a population base of more than 535,000 people. The program's goal is to involve a broad spectrum of people in discussion across social, cultural, and generational lines using significant social issues brought out in the selected text. The project also promotes literacy, and is the first "one book" project in the country to incorporate an original adaptation for the stage.

All Pikes Peak Reads 2010 Books

Lightning in His Hand: The Life Story of Nikola Tesla
by Inez Hunt & Wanetta W. Draper

The Invention of Everything Else
by Samantha Hunt

The Devil in the White City:
Murder, Magic, & Madness
at the Fair That Changed America
by Erik Larson

The Invention of Hugo Cabret:
A Novel in Words & Pictures
by Brian Selznick

Snowflake Bentley
by Jacqueline Briggs Martin

Featuring
I AM NIKOLA TESLA
by Murray Ross and Co.
performed by THEATREWORKS
University of Colorado, Colorado Springs

Lightning in His Hand

The Life Story of Nikola Tesla

by

Inez Hunt & Wanetta W. Draper

Published by
Pikes Peak Library District
with the
Nikola Tesla Museum · Belgrade
for
innovation MMX
All Pikes Peak Reads

Lightning in His Hand: The Life Story of Nikola Tesla

This publication was made possible by private funds. Interpretation of events and conclusions are those of the authors, and do not necessarily represent the opinions of Pikes Peak Library District (PPLD), PPLD Board of Trustees, or PPLD employees.

Originally published: Denver Colo., Sage Books, © 1964 Inez Hunt & Wanetta W. Draper. Reprinted with permission. New foreword and illustrations added.

Endorsed by

COLORADO HUMANITIES

Font Tesla, used on the title page, courtesy the Nikola Tesla Museum, Belgrade, Serbia. The font represents Tesla's handwriting and was based on writing samples from the museum collection. Idea: Rastko Ćirić; Font creators: Vladimir Popović, Olivera Stojadinović; Preparation and research of archive material: Ivana Zorić, Milica Kesler, Andrija Panić, Zorica Civrić

Nikola Tesla Museum, Krunska 51, Belgrade, Serbia
http://www.tesla-museum.org

Publisher's Cataloging-in-Publication Data

Hunt, Inez.
Lightning in his hand: the life story of Nikola Tesla / by Inez Hunt and Wanetta W. Draper.—
Revised ed.
p. cm. (Regional history series)
Includes bibliographical references and index.
LCCN 2010924729
ISBN 978-1-56735-279-5
1. Tesla, Nikola, 1856-1943. 2. Electrical engineers—United States—Biography. 3. Inventors-United States—Biography. 4. Colorado Springs (Colo.) I. Draper, Wanetta W. II. Series.

TK140.T4
621.3092-dc22

Regional History Series

Currently In Print

The Colorado Labor Wars: Cripple Creek 1903–1904, A Centennial Commemoration

"To Spare No Pains": Zebulon Montgomery Pike & His 1806–1807 Southwest Expedition

Doctor at Timberline: True Tales, Travails, & Triumphs of a Pioneer Colorado Physician

Legends, Labors & Loves: William Jackson Palmer, 1836–1909

Extraordinary Women of the Rocky Mountain West

Forthcoming

Doctors, Disease & Dying in the Pikes Peak Region

Rush to the Rockies: The 1859 Pikes Peak or Bust Gold Rush

El Paso County Inventors & Their 19th Century Patents

For purchasing information, contact:

Clausen Books
2131 North Weber Street
Colorado Springs, Colorado 80907
tel: (719) 471-5884, toll free: (888) 412-7717
http://www.clausenbooks.com

Acknowledgements

The Editorial Committee is grateful to the authors of *Lighting in His Hand*. We are also grateful to their children for working with the Pikes Peak Library District to republish this book as a special addition to our Regional History Series. Our thanks go to Inez Hunt's daughter, Fran Kiester, and to Wanetta W. Draper's children, Sandra Dowda, Paul Draper and Brad Draper.

Nina Kuberski digitally captured the original text for re-setting, making the reprinting a much easier task. Katie Rudolph, photo archivist, created the striking cover, and both she and Nina assisted with preparing the illustrations for this publication.

We are especially indebted to Vladimir Jelenković and the staff of the Nikola Tesla Museum, Belgrade, Serbia, for their assistance in this publication. The Museum provided several new photographs of Tesla's laboratory in Colorado Springs, helping to make the book even more relevant to the Pikes Peak Region.

Richard Marold, Tesla scholar and portrayer, deserves special recognition for undertaking the research necessary to "become" Nikola Tesla at events related to this book. Richard is well known for his talented portrayals of Winfield Scott Stratton and Franklin Delano Roosevelt, and he did not hesitate to darken his hair and devour all he could read about the intriguing man, Nikola Tesla.

The Editorial Committee

CONTENTS

Foreword • v

Preface • 1

"Little London" • 5

Tesla's Early Childhood • 9

Tesla's European Education • 21

From Budapest to America • 31

The Battle of the Currents Begins • 41

Telluride, Colorado's Contribution to Electrical World History • 57

Lighting the Columbian Exposition • 71

A Columnist's Viewpoint • 79

Niagara Falls & the Termination of the Battle of the Currents • 89

Experiments in Resonance & Teleautomatics • 95

Tesla's Laboratory in Colorado Springs • 105

Tesla Discovers Stationary Waves in the Earth • 119

Tesla's Most Dramatic Colorado Experiment • 131

Tesla's Return to New York: Construction of Wardenclyffe Tower • 141

Development of Turbines • 163

The Mystery of the Nobel Prize • 169

The Award of the Edison Medal • 177

Tesla's Involvements with the Occult • 187

Tesla's Friendship with
Robert Underwood Johnson • 205

Cumulative Honors Awarded to Tesla • 225

Tesla's Death Ray & the Projection of
His Ideas into the Future • 243

Death of Tesla • 255

Bibliography • 265

Index • 273

About the Authors

Inez Hunt (1899–1983) & Wanetta W. Draper (1917–2006) collaborated throughout their writing careers and co-authored many books, including *Ghost Trails to Ghost Towns* (1958); *Horsefeathers and Applesauce* (1959); *To Colorado's Restless Ghosts* (1960); and *Lightning in His Hand: The Life Story of Nikola Tesla* (1964), published by Sage Books. Later works included *Colorado Crazy Quilt* (1971), published by HAH publications. Much of the research for *Ghost Trails to Ghost Towns* was completed on joint family camping trips. During the writing of the text *Horsefeathers and Applesauce*, Inez and Wanetta founded the College of Western Research and Psychic Phenomena. "A fun project to show how easy a diploma mill can be established in this state." After paying $3.50 and submitting a request to Denver, the college was established.

FOREWORD

One hundred eleven years ago, on May 17, 1899, a brief but electrifying chapter in Colorado Springs history began when Nikola Tesla stepped off a Denver and Gulf Railway car and was taken by his friend Leonard Curtis to engage a suite at the Alta Vista Hotel. Although the visionary scientist's time in Colorado Springs was short, less than one year, the insights visited upon him during grand electrical experiments in his barn-like laboratory on Knob Hill profoundly affected the course of his life and ideas. Lured to Colorado Springs by a combination of its isolation, its unique, high-altitude atmospheric conditions, and Curtis's promise of free electricity, Tesla stayed only long enough to test and confirm his ideas on the wireless transmission of energy. The region's magnificent lightning storms inspired both his scientific thought and his poetic imagination.

Inez Hunt and Wanetta Draper's insightful biography of Nikola Tesla goes well beyond describing the role Colorado Springs played in Tesla's story, and presents a comprehensive portrait of the man who, perhaps more than any other, is responsible for conjuring our technological world. When Hunt and Draper, both Colorado Springs writers, first published *Lightning in His Hand* in 1964, there was ample evidence of the influence of Nikola Tesla's ideas on the "modern" world. Alternating current electricity, radio, remote control, and x-ray technology were but a few of the practical applications of Tesla's experiments. His "teleforce" weapon, dubbed the death ray, fueled post-nuclear America's Cold War paranoia and became a staple of science fiction B cinema.

Perhaps an even greater influence has been manifested in our own time, when Tesla's ruminations on wireless communication devices and instantaneous access to global information certainly have come into their own—more than 100 years after he described them. Likewise,

Tesla's visions of drawing free and limitless energy from the sun's powerful rays or the earth's inner heat, and his concept of the electric automobile, are ideas just beginning to gain acceptance and whose ultimate benefits may yet to be realized.

Although largely forgotten in Colorado Springs, Tesla's spirit still crackles through the dry Rocky Mountain air with each summer lightning storm—and his story, told so well in this book, may attune you to hear it. The Pikes Peak Library District's republication of Inez Hunt and Wanetta Draper's *Lightning in His Hand: The Life Story of Nikola Tesla* is an attempt to keep that spirit alive and pay it a deserved homage.

Paula J. Miller, Executive Director, PPLD

Dennis Daily, Assistant Manager, Special Collections, PPLD

About Pikes Peak Library District

Pikes Peak Library District (PPLD) is a nationally recognized system of public libraries serving a population of more than 547,000 in El Paso County, Colorado. With twelve facilities, online resources, and mobile library service, PPLD responds to the unique needs of individual neighborhoods and the community at large. PPLD has an employee base of 432 full and part-time staff, and utilizes roughly 1,500 volunteers. It strives to reach all members of the community, providing free and equitable access to information and an avenue for personal and community enrichment. PPLD is recognized for its commitment to diversity and community collaboration, its quality programming, and its excellent customer service.

For his enthusiastic love of history,
inspiration of the Regional History Series books,
& generous gifts of his time and talent,
we dedicate this book to

Calvin P. Otto (1930–2009)

Lightning in His Hand

Preface

The scientist has written of the material achievements of Nikola Tesla. The biographer has dealt with the vital statistics of his life. The occultist has related this man's prophecies to his own mysticism. The journalist has often played up the sensational aspects of Tesla's career. Full understanding, however, involves the combining of the many facets which compose the personality of this strange genius of electricity.

It is for this reason that we have garnered the information for this book from wide sources. We have explored sundry viewpoints; we have considered the power developments which were triggered by his wizardry. They are all a part of the Tesla story.

Especially helpful to us has been the acquisition, by the Library of Congress, of unpublished correspondence which shows a new profile of this important figure and his relationship with friends and with the great financial magnates of a fantastic era in world history.

Through correspondence and interviews with many who knew him personally, much fresh material has come to light. From the files of international journals and news media of the world, we have brushed the dust and brought to light buried bits that have strange prophetic bearings on the jet-power propulsion of the day-after-tomorrow which is already here.

We have two purposes in writing this book. First, that Tesla may not be forgotten by a forgetful world. Second, that some young scientist may be inspired by the challenge of a neglected heritage of Tesla's ideas. We ask of him to pursue some of the uncompleted dreams of this great man so that the brain of this genius may not have been short circuited by death.

Inez Hunt & Wanetta Draper
1964

Looking west down Pikes Peak Avenue, the first Antlers Hotel is pictured left of center. The Antlers burned on October 1, 1898, requiring Nikola Tesla to stay at the newly modernized Alta Vista Hotel to the north on Cascade Avenue. This photograph was taken in the vicinity of where Tesla constructed his Colorado Springs laboratory in 1899. According to *The Giles City Directory of Colorado Springs* for 1900, the "Tesla Experiment Station" was located "beyond Deaf & Blind Institute, 1 mi[le] e[ast] of P O via Pike's Peak ave." The population of Colorado Springs in 1900 was 21,085. *From Special Collections, Pikes Peak Library District, (013-616).*

"Little London"

"Little London" was asleep. The day's social whirl had drawn to a close for the English gentry and their imitators who had gained the sobriquet for Colorado Springs. Dancing slippers had been kicked off in the palatial North-end homes for the servants to put away in the morning. Corsets had been unlaced. Heavy gold watches and their vest chains and fobs had been deposited on marble-topped dressers. Kid curlers had been twisted in their nightly ritual of feminine vanity and stray combings had been dropped into the hand painted hair receivers. Lights had been turned out. Little London was asleep.

Just east of the Deaf and Blind Institute, on the empty prairie-pasture that was known as Knob Hill, a strange barn-like structure was silhouetted in the dark autumn night and the atmosphere seemed charged with that oppressive quiet which precedes storm.

If any restless wanderer of the night had happened to be trespassing in that neighborhood, he would have seen Nikola Tesla inspecting his electrical apparatus in the strange laboratory, but he would have sworn he saw the devil walking—clad in white tie and tails—tall, mysterious and eerie as he looked upward toward the copper ball on the two-hundred-foot mast.

Such a wanderer would have seen Kolman Czito with his hand tense upon the switch, waiting for Tesla's signal. He might even have heard the tall weird figure—two inches taller that night because of his heavy insulator soles—call out, "Czito, close the switch—now!" But the passerby would have heard no more before all hell broke loose, and the night belched forth flames, sulphur fumes, and rods of fire-like emanations that in seconds reached lightning proportions. He might have been deafened by the reverberations of thunder, loud as artillery battle, as the sound echoed to Cripple Creek fifteen miles over the giant ridge.

Heavy sleepers over town stirred and some may have awakened to close their windows. Such thunder might bring rain. Rain could ruin lace curtains.

But there was no more thunder. Nikola Tesla's laboratory suddenly became quiet where moments before, more power had been loosed than that generated at Niagara Falls.

Few people in Colorado Springs knew the importance of this newcomer or inquired about this unfathomable stranger who folded his long legs into the Antlers Livery buckboard each day for the jog to his laboratory. Nikola Tesla was beyond the comprehension of the average man on the street. Few knew that already to his credit was the invention of the first practical means for the use of alternating current—the invention (plus his polyphase system) that had ushered in the Power Age. They accepted the arc light on the corner with no realization of the part this man had played in its development. They were too far from Niagara Falls to be cognizant of his name on the plates of the huge machinery there. Turbines, generators, rotary magnetic fields, and oscillators formed the language remote from their daily vocabularies. Neither could they have been expected to have been clairvoyant enough to have foreseen such developments as radar, television, radio, certain concepts of cosmic force, death rays, and thermo electrons, all of which were to comprise a small part in the myriad of ideas in Tesla's world.

So, in the autumn of 1899, Colorado Springs concerned itself with culture, the price of gold, railroads to the new mining camps, Paris fashions, and the social whirl. There was scant time for the impractical dreamer who was doing tricks with electricity and homemade lightning, all of which had little meaning and was of absolutely no importance to their daily scheme of living.

Little London was asleep.

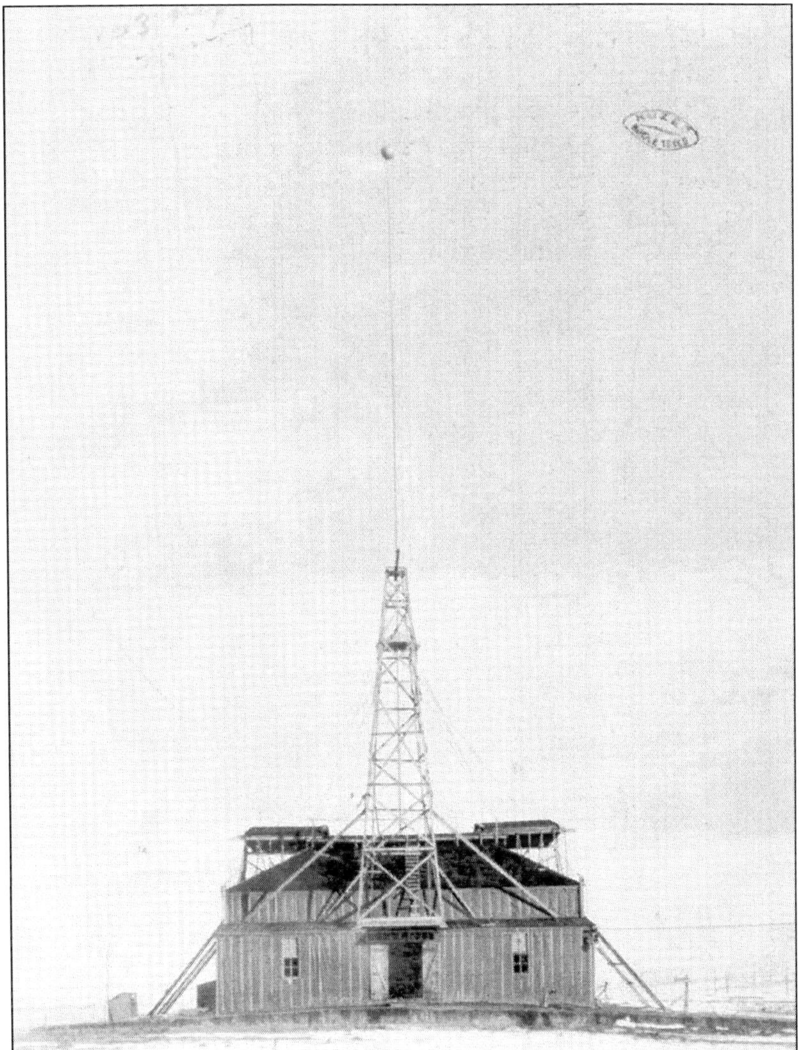

Nikola Tesla's Colorado Springs laboratory taken from the west side, late December 1899. Tesla wrote that this photograph was "taken in the morning by sunlight, fresh fallen snow on the ground." The photograph was made by Tesla and New York City photographer Dickson D. Alley. *Courtesy Nikola Tesla Museum, Belgrade, Serbia, (006-04-7).*

Smiljan, Croatia, was a small village when Nikola Tesla was born in 1856. Too small to be included on this 1860 map, Smiljan is located just 5 kilometers northwest of Gospich (now Gospic), in the lower left corner. Gratz (now Graz), where Tesla attended school is in the upper left corner, and Belgrade is in the lower right corner. *Johnson's New Illustrated Family Atlas, New York, N.Y.: Johnson and Browning, 1860.*

Tesla's Early Childhood

The world was dark at the stroke of midnight in the village of Smiljan, when Nikola Tesla was born, July 9, 1856. Actually, it was a matter of question whether it could have been July 10, for it was at that faint breath of time between one day and the next which belongs to neither.

The household of the Reverend Milutin Tesla had eagerly awaited the birth of a fourth child. The Tesla heritage was solid, with a background of scholarship, military honors, clerical dedication, and manual dexterity. The family name, formerly Draganic, is translated, Darling[1]. Through the years, the members of one branch of the family adopted the nickname of Tesla, the term used for an adz-like tool; the name also denoted a family characteristic of a protruding front tooth of that shape.

The mother came from the household of Mandich with a lineage preponderant with ministers. Djouka was her given name, which, translated, is Georgina. Although she had no formal education and never learned to read or write, she was nevertheless an intelligent and capable person. Since her mother's blindness had made it necessary for Djouka to forego school and care for the family, she developed a Homeric method of storing thousands of verses of poetry and scripture and the sagas of her country in her memory. When she married she became the capable manager of the financial affairs of her household. She accepted her place in the church life of the community. She took pride in her children. Her home reflected her efficiency, with many devices which she invented for that purpose. Exquisite needlework, for which she was noted, was proof of her nimble fingers. It has been said that when she was over sixty, "using only her fingers, she could tie three knots in an eyelash."[2]

It was from this remarkable woman that the Tesla children received their important early impressions, and it was to her that Nikola Tesla, in later life, attributed his inventive abilities. From his father he inherited a love of poetry, an interest in philosophy, a reverence for books, and—a touch of impracticability.

Although Nikola Tesla's birthplace was in Lika, Croatia, the family was of Serbian origin. In this year of 1856, Smiljan was governed by Austria-Hungary. It is now, in 1963, a part of Yugoslavia. Although Croatia was predominantly Roman Catholic, the Reverend Milutin Tesla was a priest of the Serbian Orthodox faith, and he conducted services in the little village of Smiljan, a community of about forty families.

There is legend that Nikola Tesla was born during a dazzling electrical storm and that the midwife predicted that he would be a child of storm.[3] The story credits the mother with evading the connotation, preferring to predict that her son would be a child of light. The father added his own interpretation. He was certain that his son would spread the light of the gospel.[4]

Upon Milka and Angelina fell some of the big-sister responsibilities of sharing the care of Nikola. Seven-year-old Dane may have felt the twinges of sibling rivalry. The villagers doubted if the new baby would ever equal his older brother, for whom great things had been predicted. It seemed doubtful that any family should be blessed with a second son as gifted as Dane.

When Marika, the fifth child, was born, the small white parsonage adjoining the church was fairly bursting at the seams, and Djouka's practical abilities were taxed to the limit. However, the child Nikola felt no crowding. As he grew older, his world became made up of the woods and nearby fields and hills. He peered into bird nests and became an expert with a sling shot. At one time he brought home a baby eagle, firmly convinced that he could learn from it the secret of flying. The obsession with flight was to grip him all his life. The geese overhead in Smiljan were indelible in his memory. Over a half century later he recalled these geese:

> . . . which used to rise to the clouds in the morning and return from the feeding grounds at sundown in

> battle formation, so perfect that it would have put a squadron of the best aviators of the present day to shame.[5]

It was the brook near his home which inspired him to attempt one of his first inventions, at the age of four. With the help of his brother Dane, he improvised a water wheel from a crude disk. He was firm in his directions to Dane as to the way he wanted it done. Then he constructed an axle from a twig and felt his first thrill of inventive achievement when the disk turned in the stream. Most water wheels had paddles, but this crude paddleless wheel remained in his mind and became the underlying principle, many years later, for his invention of his smooth disk turbine without buckets.[6]

His investigative turn of mind caused the family continual anxiety. He was always in trouble. He became lost and locked in an old country church long after dark, and the entire neighborhood joined in the search for him. Another time he fell into a tub of hot milk that luckily had not reached a temperature hot enough to scald him.

His imagination led him to do battle with the garden cornstalks. With a wooden sword he attacked them. Years later he recounted the escapades, which, he said,

> . . . ruined the crops and netted me several spankings from my mother. Moreover, these were not of the formal kind but the genuine article.[7]

Not the least of these childish adventures was his initial attempt at flight by means of the family umbrella when he decided to parachute from the roof of the barn. He had an intuitive feeling that by the force of his will he might become weightless and fly through the air. Failure on his trial flight battered his body and ruined the umbrella but did not defeat his spirit or determination to discover the mysteries of aeronautics.

Until the age of eight he considered his character to be weak and vacillating, and he felt he had neither courage nor strength to form a firm resolve. He suffered fears of pain and death—of darkness—of religious condemnation—of ogres, and the spirit of evil—of ghosts and unholy monsters that lurked in the dark. Perhaps it was darkness that was to be his dragon, still to be slain.

The Reverend Tesla may have been dismayed by the escapades of his disconcerting young son, but he had no doubt that life would discipline the lad for his wayward antics and make him all the better, someday, for the chosen field of the ministry.

A tragic accident took the life of twelve-year-old Dane while the family lived in Smiljan. All the community mourned with the Teslas for the loss of the brilliant child. Nikola, who had hitherto been overshadowed by his brother's advanced intellect, felt the necessity, in some way, to fill the void that had been left by Dane's death.

Shortly after this, the father was transferred to a larger church in Gospic. The new environment created a difficult adjustment for Nikola. Accustomed to roaming the hills at will, the confinement of urban life imposed a hated discipline. In his new house, he became a prisoner watching strange people through the window blinds. He described his shyness

> such that I would rather have faced a roaring lion than one of the city dudes who strolled about.[8]

He missed the pigeons, the chickens, and the sheep that had made up his pastoral world in Smiljan.

So painful was his shyness, that his father allowed him to hide away one Sunday morning in the belfry tower. He literally put his worst foot forward that day. At the close of the service, he raced down the tower steps, and, leaping from his retreat, he landed on the flowing train of the dress of the most influential parishioner in the congregation. The torn dress did not endear him to the domineering woman who held the purse strings of the church. Nor did it help public relations in the community which was accustomed to taking its cues from this important person. Some kind of redemption was called for to get back into the good graces of the social arbiters and vindicate his actions to a world already prejudiced and aware of the tradition of ministers' sons.

Such reprieve came at a celebration of the purchase of the very latest fire-fighting equipment for Gospic. A man-powered pump, which operated a hose designed to draw water from the river, was to be initiated with festivities. It was a picnic sort of holiday. City officials seized the opportunity to orate at length before a captive audience which waited more or less patiently to see the new pump in action. At last, at a given signal, sixteen men applied their muscles to the new, brightly painted machinery. Nothing happened. The vision of their tax money injudiciously spent caused the townspeople to grow tense and alarmed. The spectre of lost prestige and political censure loomed over the heads of the frustrated and helpless officials.

Young Nikola vindicated himself by one more leap—this time under more auspicious stars. Plunging into the river, he straightened the kink in the hose which had been caused by the current. The water gushed out of the nozzle. The seven-year-old hero was hoisted to the shoulders of the firemen and paraded through the town. All was forgiven. When he was asked how he knew what to do, he could not explain. His action had been intuitive. He did not know how he had known.[9]

Nikola soon became aware that there were certain characteristics which set him apart from his fellow students, his family, and the world of ordinary beings. He had started to school at a younger age than customary, but it was not just an accelerated mind that set him apart. He was conscious of certain phenomena before his eyes which other people could not see. He envisioned objects and hypothetical situations and day dreams with such reality and clarity that he was uncertain whether they did, or did not, exist.

When these visions became so real, he confided his dilemma to his sister, who tried to help him to distinguish the real from the imaginative. Nikola wrestled with his problem but he dared not discuss it with others. He felt keenly the stigma of being different. He could not bear the derisive laughter of companions and he did not want to cause embarrassment to his family.[10] But he realized that he was, literally and figuratively, a left-handed person in a right-handed world.

Often the Reverend Tesla was complimented upon the inventive ability of his son, and his friends suggested that the boy might become a scientist some day. The father was always quick to reply that Nikola was promised to the ministry. More and more, young Nikola Tesla was inwardly rebelling against the career that was going to be forced upon him. Family ties were close. He had no desire to hurt or disappoint his parents, but some compelling drive was forcing him in another direction.

When he had reached the age of fifteen, he was sent to Higher Real Gymnasium at Karlovac (Carlstadt), Croatia, to live with a relative of his father. There he contracted malaria. His aunt felt that his diet should be carefully restricted, with the result that he was hungry all of the time. Describing this part of his life he said:

> . . . She was a distinguished lady, the wife of a colonel who was an old war horse, having participated in many battles. I can never forget the three years I passed at their house. No fortress in the time of war was under more rigid discipline. I was fed like a canary bird. All the meals were of the highest quality and deliciously prepared, but short in quantity by a thousand percent. The slices of ham cut by my aunt were like tissue paper. When the colonel would put something substantial on my plate, she would snatch it away and say excitedly to him, "Be careful, Niko is very delicate." I had a voracious appetite and suffered like Tantalus, but I lived in an atmosphere of refinement and artistic taste quite unusual for those times and conditions.[11]

Although the colonel was concerned with appeasing the physical appetite of his hungry nephew, he had little patience with his mechanical contrivances or inventive aspirations. Inspired by the mechanical models in the school room, Tesla turned his attention to water turbines. He constructed many of them and took pleasure in operating them. He may have remembered the little water wheel in the Smiljan brook. His uncle could not share the joy of creating these things. There was something uncanny in the boy's ability to instill magic into wood and wire. He did not understand the lad's turn of mind. There was almost a quality of his being in league with the devil that was disturbing. So the uncle often rebuked Nikola for the time he wasted on such things.

Once Nikola opened an old book and saw a steel engraving of Niagara Falls. He was fascinated by the picture and he imagined how a great wheel might be made to turn in the tremendous power of that turbulence. Once in a day-dreaming mood, he confided to his uncle the feeling that some day he would go to America and harness the runaway waters of Niagara. Thirty years later he was able to report:

> I saw my ideas carried out at Niagara and I marveled at the mystery of the mind.[12]

Living in a military environment, young Tesla began to develop a self-imposed discipline. He began to deny himself whatever he wanted most. He learned to renounce anything which gave him pleasure. At this point in his life he was formulating a decision for his career, yet he was troubled because he knew it meant thwarting his father's ambition for a son's consecration to the ministry. He was coming to the realization that his chosen field would demand all the asceticism and sacrifice imposed upon the dedicated. His rigorous discipline set the early pattern for a selfless life of service.

As he recalled later in a biographical article:

> When I was a boy of seven or eight ... the possibilities of willpower and self-control appealed tremendously to my vivid imagination and I began to discipline myself. Had I a sweet-cake or a juicy apple which

> I was dying to eat, I would give it to another boy and go through the tortures of Tantalus, pained but satisfied. Had I some difficult task before me which was exhausting, I would attack it again and again until it was done. So I practiced day by day, from morning till night. At first it called for a vigorous mental effort directed against my disposition and desire, but as years went by the conflict lessened and finally my will and wish became identical. They are so today, and in this lies the secret of whatever I have achieved. These experiences are as intimately linked with the rotating magnetic field as if they formed an essential part of it; but for them I would never have invented the induction motor.[13]

The land surrounding the Brancovic home in Carlstadt was low and marshy. Everyone consumed enormous amounts of quinine to counteract the threat of malaria. Occasionally the river would rise and drive armies of rats landward, which would devour everything in their path. This was terrifying to some, but Tesla welcomed the invasion, proving his resourcefulness by investigating all sorts of extermination and gaining for him the title of questionable distinction, "rat-killer of the community."

School made the stay at Carlstadt bearable. When Tesla became engrossed in the study of physics, he forgot his gnawing hunger and his recurrent bouts with malaria. He was particularly stimulated by the influence of one professor who was an ingenious man and given to demonstrating apparatus of his own construction. Among these inventions was a device in the shape of a freely rotatable bulb with tin-foil coatings which was made to spin rapidly when connected with a static machine. Tesla was overcome with an intensity of feeling of precognition as he witnessed these exhibitions. Every impression, he said, produced a thousand echoes in his mind. He sensed a wonderful force and a longing for total submission to it.

At last when his course was completed and he obtained his certificate of maturity, the day for which he had lived arrived—the day to go home to Gospic! But also came the message from his parents that he was not to return because of a cholera epidemic. The plague, deadly as it was, was not enough to stop the homesick boy, and to Gospic he went. The townspeople believed that the scourge was air-borne, so Tesla found the village in a smoke pall. Pungent shrubs were burning all over town to purify the air. Meanwhile the villagers continued to drink the infected water and to die in heaps. The smell of death was everywhere.

Young Tesla was in no condition to ward off the disease. He was thin and undernourished, and his entire system was ready prey for the virulent infection. Although he survived the crisis, he was confined to his bed for nine months with scarcely the ability to move. He found his energy completely depleted. He faced death with no particular will to live. His lethargy was augmented by the conflict between his growing compulsion toward a life of science and his father's determination for his early entry into the ministry. "Perhaps," said Tesla when he was being urged to try to recover, "perhaps, I might get well if I could study engineering."

The father came to the realization that the boy must have every incentive to live. He had lost one son and dared not lose another. In this moment of defeat he sacrificed his dream for his son and promised Nikola that if he would live he should have the best technical training available. Tesla often said that this reprieve lifted a heavy weight from his mind, and there are many who attribute his recovery to this. But Tesla also credited his cure to a marvelous folk remedy administered by an old lady.

> There was no force of suggestion of the mysterious influence about it. Such means would have had no effect whatever on me for I was a firm believer in natural laws. The remedy was purely medicinal, heroic, if not desperate; but it worked and in one year of mountain climbing and forest life I was fit for the most hazardous bodily exertion.[14]

At another time he was more explicit about the cholera medicine, saying:

> . . . the relief would have come too late had it not been for a marvelous cure brought about by a bitter decoction of a peculiar bean. I came to life like another Lazarus, to the utter amazement of everybody.[15]

During his convalescent period he read incessantly. The public library allowed him access to books which had been long neglected.

They entrusted them to him for classification of the works and preparation of catalogues. One day he was handed a few volumes of new literature unlike anything he had ever read before. He found these books so captivating

> . . . as to make me utterly forget my hopeless state. They were the earlier works of Mark Twain and to them might have been due the miraculous recovery which followed. Twenty five years later, when I met Mr. Clements (sic) and we formed a friendship between us, I told him of the experience and was amazed to see that great man of laughter burst into tears.[16]

Villagers may have thought the young man wasted his convalescent period out of doors day dreaming, but they were unaware of the quality of his dreams and the force of his mental activity. Tesla had ideas which he feared to divulge. He was working out engineering feats in his mind, but there was no one with whom he could share his speculations. He was envisioning a submarine tube through which letters and packages might be conveyed across the seas. He mentally constructed pumping plants intended to force water through the tube. He was daring in his assumptions of the velocity of the water. He envisioned a ring around the equator which would float freely and could be arrested in its spinning action. All these he filed away in his mind for future reference. His life had been spared for a reason. Somewhere within his dreams the reason lay hidden.

Notes

1. John J. O'Neill, *Prodigal Genius,* (New York, N.Y.: Ives Washburn, Inc., 1944), 13.

2. Ibid., 10.

3. Helen B. Walters, *Nikola Tesla,* (New York, N.Y.: Thomas Y. Crowell Co., 1961), 3.

4. Ibid., 9.

5. Nikola Tesla, "My Inventions," *Electrical Experimenter,* March 1919, 843. O'Neill, *Prodigal Genius,* 20.

6. O'Neill, *Prodigal Genius,* 20.

7. Tesla, "My Inventions," March 1919, 843.

8. Ibid., 843.

9. Walters, *Nikola Tesla,* 39.

10. Ibid., 24.

11. Nikola Tesla, "My Inventions," in *Electrical Experimenter,* April 1919, 865.

12. Nikola Tesla, "My Inventions," March 1919, 843.

13. *Nikola Tesla, 1856 – 1943, Lectures, Patents and Articles,* (Beograd, Yugoslavia: Nikola Tesla Museum, 1956), A-196.

14. Ibid.

15. Tesla, "My Inventions," April 1919, 865.

16. Ibid., 864-865.

Tesla's birthplace in Smiljan, Lika, Croatia, and the church where his Serbian Othodox Christian father, Reverend Milutin Tesla, preached. *Courtesy of Fran Kiester.*

Nikola Tesla in his 20s. *Henri Dupont studio photograph from the Kenneth M. Swezey Papers, Archives Center, National Museum of American History, Smithsonian Institution (AC0047-0000006).*

Tesla's European Education

Nineteen-year-old Nikola Tesla had no regrets when he packed his belongings to take his first step into higher education. The Polytechnic School at Graz, in a section then known as Styria, was selected by Tesla's father because it was one of the oldest and best reputed institutions. Due to his father's teaching, Nikola's previous training was above the average, and he had absorbed knowledge of several languages and a good backlog of general information. For the first time, he was allowed the freedom of choice of academic subjects. The only difficulty was that the feast of learning spread before him, combined with his ravenous appetite for knowledge, tempted him to the impossible.

In a fever of gratitude to his parents for the opportunity to fulfill his chosen destiny, he promised himself to surprise them by a superhuman display of scholarship. He dragged his long legs out of the warm bed at three o'clock each morning to study, and closed his books and blew out the candle each night at eleven. Sundays and holidays were no exception. He was driven by an unexplained urgency. It was not surprising that he passed all nine of his examinations with top honors.

Armed with triumph and a perfect record, he went home for a short rest. Strangely enough, his father made light of Nikola's hard-won honors. This apparent lack of appreciation for his grueling efforts to please almost discouraged the sensitive lad. He did not know until years later that the professors had written to the Reverend Tesla saying that unless Nikola was taken away from his studies, he would kill himself through overwork.

There was still the great gap between the religious tenets acquired from his parents and his own compulsion toward scientific truth. He pondered long on his conversations with his father when they had discussed the widening rift between their two worlds; Nikola had tried to explain, "You say you love people. I don't. It is mankind that I love."[1]

During Nikola's school years his leisure time could not be spent in the outdoors which he enjoyed. Instead, he crowded his lungs over a library desk, consuming a hundred volumes of Voltaire, not due to any devotion to Voltaire, but rather to a mania for finishing everything which he started.

When he returned the second year to the Institute, he began with a developing maturity to tailor his courses to fit his time and to concentrate his energies, interests, and physical capacity. This time he enrolled only in physics, mathematics, and mechanics. He was already singled out by his instructors as having more than promising potential. Among his most challenging professors were Professor Rogner, who taught arithmetical subjects and geometry, Dr. Allé who taught integral calculus and kept the boy hours after classroom period giving him problems to solve. Also there was Professor Jacob Poeschl, who held the chair of theoretical and experimental physics. Tesla described Poeschl as having enormous feet and hands like the paws of a bear but all his experiments were skillfully performed with clock-like precision and without a miss.[2]

The students held him in great respect but secretly ridiculed his personal idiosyncrasies and repeated the rumor that he had worn the same suit for twenty years. It was in one of this professor's classes that Nikola Tesla was stung with ridicule in an incident that was to affect the entire world.

The acquisition of a Gramme Dynamo was the inspiration for an excited class gathering in the physics laboratory. Already the scientific world was agog with the excitement generated by the inventions of the young American, Thomas A. Edison. Here was concrete evidence of the drama to come and the light that was breaking over the world. At any moment night might be turned into day. The world was at the dawn of great discoveries and power potentials.

The innocent appearing dynamo which so stirred the students' excitement had the "horseshoe form of a laminated field magnet and a wire-wound armature with a commutator."[3] As Poeschl demonstrated the machine, the brushes sparked badly. Tesla, the young perfectionist, was quick to point out the flaws and observed that it might be possible to operate it without the commutator. In an illuminating flash of understanding, it came to Tesla that the solution was to involve an alternating flow of current, thus eliminating the necessity for commutators. This was not an entirely new idea; it was one which had never found practical application. But in this moment the embryonic solution burst upon him with such Saul-visioned intensity that he threw caution to the winds and divulged his convictions to the astonished professor and his scoffing classmates.[4]

With Edison's discoveries in the field of electrical distribution by direct current, such heresy was shocking and unthinkable. With Tesla's criticism the classroom was stunned into silence by such blasphemy from a fellow student.

Professor Poeschl, unaccustomed to such rank insubordination on the part of his worshipful class, was not about to yield his chair to a neophyte. He was quick to inform them that Herr Tesla's suggestions were impossible, impractical, and completely out of order—that such a scheme would be the equivalent of perpetual motion.

For a time Tesla smarted under the sting, and his own clear conviction wavered momentarily; then suddenly he became certain that he was right and his zeal took on the fire of truth. Again his eyes were playing tricks on him: as in his early youth, he had been perplexed and disturbed by his thoughts appearing before him as a vision, so now he saw constantly varied solutions to the problem that was perplexing him. The images he saw comprised motors and generators and were to him real and tangible.[5] All his remaining term in Graz was plagued and absorbed with this problem that at times seemed unsolvable.

It was at this period of his life that he became conscious of the fact that he was overworked and needed some form of relaxation. His first year at Graz had been one of intellectual gluttony. His second was one of intense concentration in his own fields of endeavor. He became concerned for his lack of conformity and realized his need for recreation. Therefore, he decided to join the boys in a gambling venture. With beginner's luck and perhaps the advantage of inherent card sense and a mathematical mind, he became very adept at cards, chess, and billiards. Spurred on by his gains, he plunged more and more, but he played only for recreation and always returned the money that he won. But one night when Lady Luck was late, he lost all he had, his tuition money included. The boys failed to reciprocate in the gentlemanly manner with which he had returned their losses. In his dilemma, he realized he could not go to his father, who would bitterly oppose such waste of time and money. Frequently, in discussing gambling with his father, Nikola had said, "I can stop whenever I please, but is it worthwhile to give up that which I would purchase with the joys of paradise?"[6] Such remarks evoked anger and contempt from the father. Therefore, he turned to his mother, who held a different philosophy. She understood the character and weaknesses of men and felt that their salvation could only be brought about by their own efforts. When Tesla went to her with the preposterous request for more money to recoup his gambling losses, she gave him a roll of bills and said, "Go and enjoy yourself. The sooner you lose all we possess the better it will be. I know that you will get over it."[7]

Graz, Styria, Austria, view from the Schlossberg hill, between ca. 1890 and ca. 1900. *Detroit Publishing Co. photograph, Library of Congress Prints and Photographs Division Washington, D.C., (LC-DIG-ppmsc-09577).*

With the investment of the last of the family savings, this was not a game of chance nor a game of relaxation. This time he played for keeps, to the astonishment of the boys who were so sure he would always give back his winnings. But he quit the game, knowing it was his last. With a return to self-discipline he conquered his mania for gambling. Later he said:

> I only regretted that it had not been a hundred times as strong. I not only vanquished but tore it from my heart so as not to leave even a trace of desire. Ever since that time I have been as indifferent to any form of gambling as to picking teeth.[8]

The Reverend Tesla lived only long enough to see the boy graduate from Polytechnic. In 1880, Nikola went to Prague (then in Bohemia), carrying out his father's wish for him to complete his education at the University. The persistent image of the "impossible" alternating current motor continued turning in his mind. It is difficult to say whether it was in his subconscious mind or in the vision which appeared in front of his retina, but now with eyes that peered into the invisible he began actually to detach the commutator from his dream-machine. Occasionally he returned to the practical world, realizing the cost of his education and the sacrifice it entailed from his mother, who was now a widow with other children to provide for. The American telephone had just reached Europe and the system was being installed in Budapest. This seemed an ideal employment to bolster his economic situation and he went to work.

In Budapest, Tesla was beginning to feel the need to be on his own, but budget matters were not his forte. Money was for metals and batteries and wire. These were necessities. Clothes were luxuries which he enjoyed, but they must take their proper place. On one occasion a festivity brought him up short with a realization of the inadequacy of his wardrobe, which consisted of one suit. It occurred to him that by ripping his trousers and turning them they would look practically new, but he had overestimated his ability as a tailor and although he spent the night in this endeavor, he was unable to appear in public in time for the party.[9]

During his employment with the American Telephone Company in Budapest, his inventive abilities came to the fore and while there he made a telephone repeater which became the ancestor of today's loud speaker. This, which might have brought him millions, he failed to patent.[10]

It was in Budapest that Tesla suffered a complete breakdown which may have been a nervous disorder. Although he always admitted that he had super-sensitive sight and hearing, it seemed that during this illness all of his senses reacted with acceleration. He could remember as a boy discerning objects at a distance when no others could see them, and he recalled hearing faint crackling sounds of fire in neighbors' homes when the residents had not even been disturbed in their sleep. During this illness in Budapest, he could hear the ticking of a watch with three rooms between him and the timepiece. He said that a fly, lighting on a table, would cause a dull thud in his ear. A carriage passing at a distance of a few miles fairly shook his body. The whistle of a locomotive twenty or thirty miles distant caused vibration of the furniture so strong that the pain, to him, was unbearable. The ground under his feet trembled continuously. All of this made it necessary to support his bed with rubber cushions so he could rest. The sound of voices, which he eventually analyzed as coming from roaring noises and vibrations, were frightening. He became certain that the sun's rays fell with stunning force upon his brain.

In the days of convalescence he had to summon all of his willpower to pass under a bridge or other structure because of a feeling of crushing pressure. He became tortured with a radar-sensitivity similar to that of bats, which made him conscious of objects twelve feet away in total darkness. These psychic manifestations were accompanied by physical stress. His pulse sometimes reached 260 beats a minute and the tissues of his body caused an agony of twitching and tremors.

A renowned physician gave him a sedative of bromide of potassium—it was all the doctor could suggest for the malady which he described as unique and incurable.

Tesla's own determination to live and his growing realization and anticipation of an unfilled purpose in his life helped him to recover. The problem which he had vowed to solve was coming close to realization and to outward expression.

It was a former classmate, Antal Szigety, who helped him to recover from this serious illness. Szigety was a young athlete of unusual appearance. A big head with a lump on one side and a sallow complexion combined with the body of Apollo gave him a startling appearance. He emphasized to Tesla the necessity for athletic training and persuaded him to exercise each day.[11]

One evening at sunset Tesla was walking with Szigety in the city park. It was the kind of hour for reciting poetry and Tesla knew whole books of it by heart, word for word. One of these was Goethe's *Faust.* He recalled the passage:

> Sie rücht und welcht, der Tag ist überlebt,
> Dort eilt sie hin und fördert neues Leben.
> Oh, dass kein Flügel mich vom Boden hebt
> Ich nach und immer nach zu streben!
>
> Ein schöner Traum indessen sie entweicht.
> Ach, zu des Geistes Flügeln wird so leicht
> Kein körperlicher Flügel sich gesellen!
>
> (The glow retreats, done is the day of toil:
> It yonder hastes, new fields of life exploring:
> Ah, that no wing can lift me from the soil,
> Upon its track to follow, follow soaring!
>
> A glorious dream! though now the glories fade.
> Alas! the wings that lift the mind no aid
> Of wings to lift the body can bequeath me.)

As he walked toward the sunset quoting these words, the idea came like a flash of lightning and the solution to the problem of alternating current motors appeared before him as revelation. He stood as a man in a trance, trying to explain the vision to his friend. He seized a stick, drew diagrams, the same diagrams he was to use six years later in his address before the American Institute of Electrical Engineers. He poured out his ideas to his amazed friend, and now Szigety understood perfectly the principles which had never been applied.[12] The images which appeared before Tesla seemed as sharp and as clear and as solid as metal and stone. The principle of the rotating magnetic field was clear to him. In that moment, a world revolution in electrical science was born.

At last the dream that had tortured him had become reality, and to him it was the most glorious moment of his experience. The moment for which he would gladly have died, had come, but now there was reason to live. Later he explained, "Pygmalion seeing his statue coming to life could not have been more deeply moved."[13]

Notes

1. Arthur J. Beckhard, *Electrical Genius*, (New York, N.Y.: Julian Messner, Inc., 1959), 57-58.
2. Nikola Tesla, "My Inventions," *Electrical Experimenter*, April 1919, 907.
3. Ibid.
4. John J. O'Neill, *Prodigal Genius*, (New York, N.Y.: Ives Washburn, Inc., 1944), 41.
5. Tesla, "My Inventions," April 1919, 907.
6. Nikola Tesla, "My Inventions," *Electrical Experimenter*, February 1919, 746.
7. Ibid.
8. Ibid., 747.
9. Thomas Commerford Martin, "Nikola Tesla," *Century Magazine*, February 1894, 582-585.
10. Helen B. Walters, *Nikola Tesla*, (New York, N.Y.: Thomas Y. Crowell, Co., 1961), 69.
11. *Nikola Tesla, 1856 – 1943, Lectures, Patents and Articles*, (Beograd, Yugoslavia: Nikola Tesla Museum, 1956), A-198.
12. Ibid.
13. Tesla, "My Inventions," April 1919, 909.

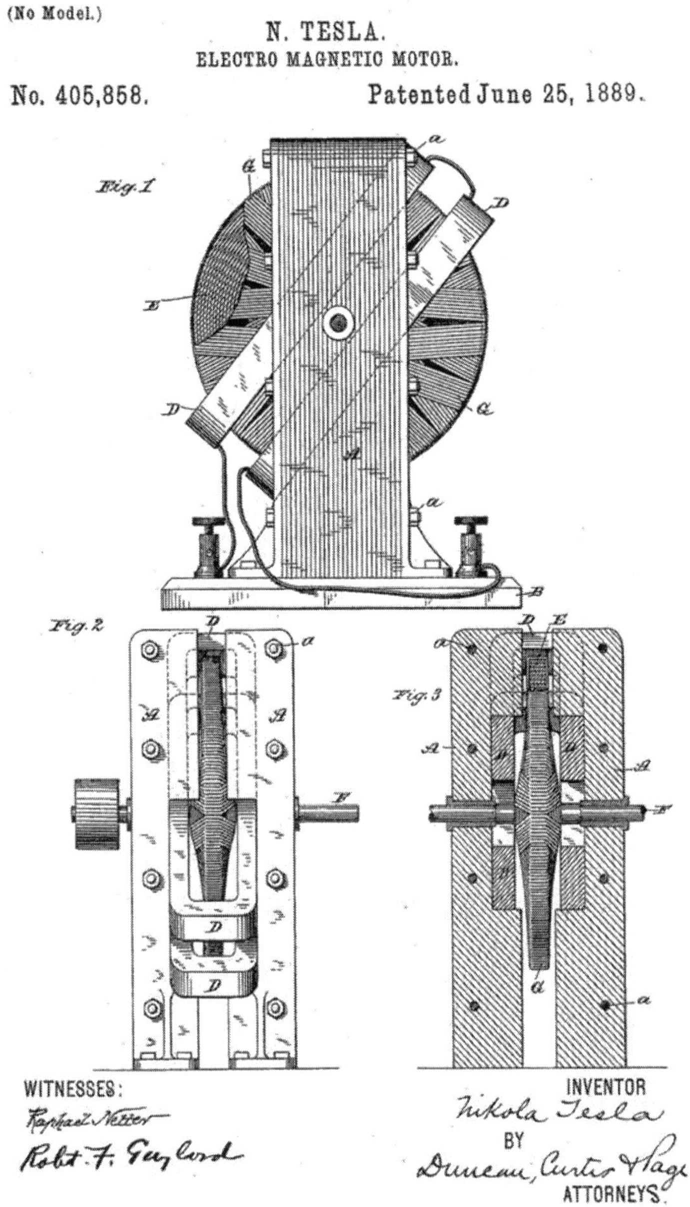

Patent diagram for Nikola Tesla's electro magnetic motor, patented June 25, 1889.

Nikola Tesla traveled to the United States aboard the *S.S. City of Richmond*, arriving in New York City in June 1884. *Sarony studio photograph, New York City, from Kenneth M. Swezey Papers, Archives Center, National Museum of American History, Smithsonian Institution (01004702).*

From Budapest to America

It was Mr. Tivadar Puskas, owner of the Telephone Exchange of Budapest, who opened the first door for the young inventor. Upon disposing of his business in Budapest, in 1883, Mr. Puskas offered Tesla an advancement in position in Paris with the Societe Continentale Edison of France. Who could resist the lure of Paris? Tesla said:

> I can never forget the deep impression that magic city produced on my mind. For several days after my arrival I roamed through the streets in utter bewilderment of the new spectacle. The attractions were many and irresistible, but, alas the income was spent as soon as received. When Mr. Puskas asked me how I was getting along in the new sphere, I described the situation accurately in the statement that "the last twenty-nine days of the month are the toughest!"[1]

Young Tesla plunged into Paris life voraciously. Rain or shine he left his living quarters on the Boulevard St. Marcel, made his way to the bathing house on the Seine for his morning plunge. Although he had lost himself in a new world, some old patterns remained. Invariably he would loop the circuit twenty-seven times; his old compulsion of numbers divisible by three regulated his behavior. After his swim he would walk for an hour to reach Ivry-sur-Seine, where he would have a woodchopper's breakfast. Here the company's factory was his outlet for work that was pure joy. A part of his time was occupied "cracking the hard nuts" for the manager of the works, the genial black-bearded young Englishman, Charles Batchellor, who was an intimate friend of Thomas A. Edison.[2]

Batchellor had originally come to Edison from the Clark Sewing Thread Mills in Newark. He had decided to remain in the United States and on having heard of Edison's electrical equipment factory, he applied for a position and was promptly employed when he demonstrated his manual skill and his ability as a mechanical draftsman. He was part of the original crew employed by the twenty-four-year-old Edison.[3]

It was only natural that Batchellor should be selected to launch the manufacturing unit at Ivry-sur-Seine. It was here that many European technicians learned to install central stations in many of the large cities on the continent.[4]

The laboratory itself was a magnet for Tesla where he met European technicians and other engineers who were to make important contributions in the years to come. There was the added incentive of prestige to be gained by working for Edison, who had achieved the highest official honors at the Paris Exposition and had been awarded the Legion of Honor. All this had made a powerful impression on Europe.

Here Tesla was thrown into contact with his first Americans and their different outlook. They fairly fell in love with him for his proficiency—in billiards! With them he discussed his dream of alternating current solution of electrical problems and one of them suggested that a stock company be formed. This was a completely new idea to Tesla; he knew only that it was an American way of doing things. But nothing came of it. No one in Paris was visualizing a revolutionary upheaval in the great Edison system, particularly one based on an untried scheme of a brilliant but hare-brained inventor, not yet dry behind the ears.

So he was sent by the company from one place to another in France and Germany. He was particularly equipped to go into Germany because of his facility with the language and his natural diplomacy combined with his engineering efficiency. Seeing the need for automatic regulators, he turned his inventive powers to the advantage of the delighted officials of the Edison Company.

Trouble arose in Strassburg, Alsace, with defective wiring. At some opening ceremonies a short circuit caused a wire to be blown out and a wall to collapse. Old Emperor William I was on hand to view the festivities and was an eyewitness to the fiasco. There was a new strain between German and French relations. Because of the flaw in the wiring, the Germans refused to accept the plant, leaving the French in an embarrassing financial position. Tesla was sent as trouble-shooter and peacemaker.[5]

In Strassburg were a number of men who subsequently achieved fame in scientific fields. They were drawn naturally to the young Serbian genius. Later, Tesla reminisced, "There were bacteria of greatness in that old town." And then modestly added, "Others caught the disease, but I escaped!"[6]

After he had cleared his desk of correspondence and had sat in on conferences and completed the necessary repairs which had occupied him day and night, he rented a shop across from the railroad station and unloaded some of his materials that he had brought from Paris. He began to produce the motor which had been in his mind and which he had envisioned on that memorable occasion in the Budapest park. It was summer before he could actually complete his project and could see the rotation of alternating current of different phases, without sliding contacts or commutator. The old dream which had caused him so much difficulty in Professor Jacob Poeschl's class was at last a solid reality which could be proved to men who could not share visions.

There are those who say that his friend Antal Szigety was invited when the first switch was thrown in that Strassburg shop and that the occasion was one for white tie and tails, and a wagered dinner as to the success of the new motor—a dinner for which Szigety was glad to pay.[7] There are others who believe that it could have been Charles Batchellor who was the delighted host on the occasion.

But it is a substantiated fact that the mayor of the city, Mr. Bauzin, who was devoted to the young inventor, was greatly impressed with the achievement and attempted to gain financial backing for the project; but all eyes were still focused on Edison and there was no response. Tesla's achievement was ignored and he was discouraged and embarrassed. There were those who had urged him to secrecy about his invention, but Tesla felt their warnings were pointless inasmuch as he couldn't even give it away. The mayor and Tesla drowned their disappointment by digging up a good-sized allotment of St. Estephe of 1801 which Mayor Bauzin had carefully concealed during a German invasion. It seemed the logical time to consume that beverage—in quantity.

Mr. Bauzin had urged Tesla to go back to Paris where there were more opportunities, but it was spring of 1884 before all the petty details were concluded and the Strassburg plant was formally accepted so he could return. The substantial sum which had been promised him for the Strassburg work for the Edison Company never materialized. He was never paid, and the three officials to whom he appealed tossed him from one to another in what became a *circulus viciosus.*

But when he returned to Paris and confided his dilemma to the manager of the Edison plant, Mr. Batchellor urged him to go to America with a view of redesigning Edison machines. Both of them were convinced that America was the land of golden promises. Tesla had yet to learn it could also be the land of broken promises.

His decision was made in haste. With the propensity for absent-mindedness ascribed to geniuses, he managed to be robbed of his luggage, his tickets, and his money before he reached the steamship. Because of his photographic memory, he was able to recall the number on his steamship ticket and when no one claimed the reservation he was allowed to make the journey to New York. The voyage gave him plenty of time to concoct futile schemes for recouping his funds. He even sat in the stern of the boat watching the wealthy passengers and hoping some millionaire might lean too far over the rail and give the impoverished Tesla an opportunity for rescue and reward. But apparently the only rail they approached was the brass one at the bar.

Tesla arrived in America with four cents in his pocket, a book of his own poems, a scientific article, and a package of calculations relating to his plans for a flying machine. The only thing left him of any immediate value was a letter of recommendation from Batchellor to Edison, stating, "I know two great men and you are one of them; the other is this young man."[8]

If he had expected to be transported into the land of dreams, the opposite was true for he had come to a land of bleak reality. He had come from a locale rich in old-world beauty and tradition. The new country was raw, rough, and machined. He felt that it was a century behind Europe in civilization. Little did he sense that his own inventions would set it a century ahead.

Lacking the carfare to reach a friend of his who was living in New York, he walked past a shop where he saw a man working on an electrical machine. Tesla stopped in his tracks. The motor was a familiar object in a strange land. It happened that the man working on the machine was ready to give up the repair and seized upon Tesla's offer of help. When the motor was running like new, the man handed Tesla a twenty-dollar bill. Twenty dollars in less than a few hours—America was already truly proving to be a land of promise!

Some time later, when Tesla was able to reach Edison with his glowing letter of introduction, he was amazed at this wonderful man "who without early advantage and scientific training had accomplished so much."[9] By contrast, Tesla was proficient in a dozen languages, literature and art. Both men were prodigious readers, but they were worlds apart in formal education. For a moment, Tesla wondered if all of his years of school had been squandered when Edison could achieve so much without. But Tesla realized later that his period of formal training and discipline were the best assets he could have had.

Edison, also, was sizing up the young applicant with a certain amount of resentment for this very backlog of scholastic achievement and the old world culture. Nevertheless, Edison could not afford to overlook the young immigrant's abilities and soon placed trust in him.

Tesla was brimming over with enthusiasm for his alternating current motor which he was eager to share with his new employer. Edison bluntly refused to listen and told the young upstart that his plant was doing all right as it was.

But Edison's confidence was strengthened during an incident when the *S.S. Oregon,* the fastest passenger steamer of the time, had both its lighting machines disabled and its sailing time delayed. Edison was perplexed with the problem, for the dynamos had to be repaired on board the ship. He assigned Tesla to the task. The dynamos were in such bad condition that the young man spent the entire night on their repair, realizing that the deadline for sailing time must be met. He enlisted the help of the crew and by five o'clock in the morning had the machinery in good working order. As he went home along Fifth Avenue, he encountered Edison and Batchellor and a few others who were going to their homes to retire. With Edison's acute capacity for sarcasm, he greeted Tesla with, "Here is our Parisian running around at night." When Tesla explained that he was coming home from the *Oregon* and that both machines were running, Edison walked away, but Tesla overheard him remark to Batchellor, "Batchellor, this is a damned good man."[10]

Edison was a taskmaster. He expected as much of his men as he did of himself, and he required Tesla to work from 10:30 a.m. until five o'clock the next morning without a day's exception. There was no time-and-a-half and his only reward was Edison's remark, "I have had many hard-working assistants but you take the cake."[11]

Edison was a robust, compelling genius of diametrically opposed temperament to Tesla's, of a mercurial temper ranging from joking to sullenness and anger. His patience applied only to his labor, and he had no tolerance for those about him who were not keen. "He was a hard worker in his own way," and it was also said that the pot of his ideas simmered slowly and he constantly lifted the lid to look in.[12]

Tesla defined Edison's peculiar methods by saying: "If Edison had a needle to find in a haystack, he would proceed at once with the diligence of the bee to examine straw after straw until he found the object of his search." Such technique seemed criminally slow in time wasted. Tesla went on to say: "I was a sorry witness of such doings, knowing that a little theory and calculation would have saved him ninety percent of his labor."[13]

The young immigrant had no understanding of the sadistic jokes with which Edison often amused himself at the expense of others. For instance, when Tesla inquired admiringly the source of Edison's phenomenal energy, he was told that it came from a daily breakfast of Welsh rabbit. Tesla, in spite of a protesting stomach, gave it a fair trial, to his own discomfort.[14]

Edison maintained that in order to invent, one's system must be out of order and he insisted that there is nothing that will do it like good old-fashioned American pie. It is possible that his suggestion of Welsh rabbit was not entirely malicious, but that he merely felt it was one sure way to keep Tesla's digestion in turmoil and thus increase his inventive abilities![15]

As for his own diet, Edison's likes and dislikes were strong with an insistence on constant variety and change. Much as he advocated the daily Welsh rabbit and apple pie for his workers, he himself would brook no monotony and no repetition in the menu.

Edison had no knowledge of Tesla's background, nor did he care. He did not know where Tesla's birthplace was; perhaps it might have been some uncivilized area of the world, for he once shocked the young employee by asking if he had ever eaten human flesh.[16] To Edison, here was an enigmatic bohunk who was entitled to small consideration except in whatever work he might contribute to the factory.

But there were others who were impressed with the young Serbian's accomplishments. Among these was Ernest Oborne, an employee of the East Orange Telephone Company. Oborne was ten years younger than Tesla, but he shared a common interest in mathematics and in the new electrical developments. Often he bicycled to meet Tesla, to save his scarce transportation money.[17]

Although Tesla had an antipathy toward the use of direct current motors, he worked to improve Edison's dynamos. He was sure he could increase the output, lower the cost, and decrease the maintenance. Edison replied, "If you can do this, young man, it will be worth $50,000 to you."[18] This would mean the realization of a laboratory for Tesla and the means for a life of scientific exploration. This was what he had visioned as the meaning of America's golden promise. He set to work harder than ever, driving himself beyond his endurance, and as a result came up with the design of twenty-four different types of standard machines, short cores, and uniform patterns which were to replace the old ones.[19]

Edison was delighted with the results, but there was no $50,000 in Tesla's pay envelope and after some time, Tesla approached him for the money. It is said that Edison replied, "Tesla, you don't understand our American humor." Tesla didn't.

All the faith and anticipation that had gone into the past grueling months had become ashes and his disappointment was compounded with disillusionment. Years later, he simply stated, "I was the victim of a practical joke." There are some who say he immediately turned in his resignation. There is one writer who says he "tipped his hat to Edison as he left the laboratory."[20] It is possible that this is true. Tesla was a gentleman under any circumstances.

Notes

1. Nikola Tesla, "My Inventions," *Electrical Experimenter*, May 1919, 16.
2. Ibid.
3. Matthew Josephson, *Thomas A. Edison*, (New York, N.Y.: McGraw Hill, 1959), 87-88.
4. Ibid., 216.
5. John J. O'Neill, *Prodigal Genius*, (New York, N.Y.: Ives, Washburn, Inc., 1944), 54.
6. Tesla, "My Inventions," 17.
7. Helen B. Walters, *Nikola Tesla*, (New York, N.Y.: Thomas Y. Crowell, Co., 1961), 77-79.
8. O'Neill, *Prodigal Genius*, 60.
9. Tesla, "My Inventions," 664.
10. Ibid.
11. Ibid., 64-65.
12. Bailey Milliard, "Twelve Great Scientists, VII, Thomas Alva Edison," *Technical World*, October 1914, vol. 22, 278-285.
13. Josephson, *Thomas A. Edison*, 87-88.
14. Thomas Commerford Martin, "Nikola Tesla," *Century Magazine*, February 1894, 582-585.
15. Robert Underwood Johnson, *Remembered Yesterdays*, (Boston, Mass.: Little, Brown & Co., 1923), 116.
16. Martin, "Nikola Tesla."
17. Interview with Harry Oborne.
18. Tesla, "My Inventions," 65.
19. O'Neill, *Prodigal Genius*, 64.
20. Arthur J. Beckhard, *Electrical Genius*, (New York, N.Y.: Julian Messner, Inc., 1959), 117.

Thomas Alva Edison, 30 years old, pictured with his new Edison Cylinder Phonograph, 1877. Edison invented this recording and playing device while working on a telegraph transcribing machine. A famous Edison quote states, "Just because something doesn't do what you planned it to do doesn't mean it's useless." *Levin C. Handy photograph, Library of Congress Prints and Photographs Division Washington, D.C., (LC-DIG-cwpbh-04043).*

Nikola Tesla, ca. 1904. *From Kenneth M. Swezey Papers, Archives Center, National Museum of American History, Smithsonian Institution (01004701).*

The Battle of the Currents Begins

The graph of Tesla's next two years took sudden drops and sporadic ascents. There was the promising mirage of arc light developments, plus the renewed dreams for his alternating current motor. During this time, a group of promoters who were interested in arc light projects offered him the security of a laboratory of his own and a company under his own name. He was soon to find that this group was willing to exploit his talents but that his pay would be chiefly a name plate on his door and engraved certificates which did not even grant him voting power. He was to learn that these new associates were interested only in the arc light, and any hope for the promotion of his alternating current motor was strictly a delusion.

It is a sharp descent from the heights of Cloud Nine to the bottom of a New York sewer ditch, but this is where Tesla was to be found in this period of his life—a period which afterward he would like to have forgotten forever. His failure to communicate his dream to Edison or to any of the promoters brought on an acute depression possible only to such a volatile genius. He momentarily abandoned all hope of succeeding in the world which had seemed to defraud him. There was little work opportunity in his field; it was a time of national economic stress. When he was offered a job as ditch digger, he took it and was grateful for the two dollars per day in his pay envelope. For several months he grasped at whatever jobs he could find to supplement his earnings.

If there was a fortunate aspect to this experience, or a finger of fate, it was the coincidence that his foreman was also a man working at a job beneath his abilities. This man was capable of seeing the wasted potential of Tesla in menial labor. Because of this foreman's interest, Tesla was introduced to A. K. Brown of the Western Union Telegraph Company. Brown found a friend to help him provide the means for a laboratory so that Tesla might build his motors and work out his ideas as he saw fit.[1]

It was through Brown that George Westinghouse, Jr., head of Westinghouse Electric Corporation and an inventor of numerous electrical devices, was requested to inspect Tesla's inventions. Westinghouse was a daring young electrical pioneer, the same age as Edison and of the same reckless spirit. He was already well known for his air-brake inventions for trains. He was a man of imagination but also a practical business man. Westinghouse had no problem of protecting a competing invention such as Edison had in his incandescent lamp, and he did not hesitate to consider Tesla's alternating current motor. Here, at last, was the first listening ear Tesla had found.

Westinghouse was quick to recognize in Tesla "an inspired genius, into whose mind inventions sprang, as the conception of a great picture projects itself upon the imagination of an artist."[2]

Westinghouse had already played with the idea of the alternating current transformer when he purchased the American rights, patented in England in 1883, of Lucien Gaulard and John Gibbs. For these rights he had paid $50,000. But experiments with these ideas in his Pittsburgh laboratory had revealed faults and proved unsatisfactory. Further trial in Great Barrington, Massachusetts, in 1886, tested the workability of the transformer system and this town became the first American community to be lighted by alternating current transformers. This experiment at Great Barrington was a factor in the foundation of the Westinghouse Corporation.[3] Thus had already begun the rumble of artillery that heralded the "Battle of the Currents."

On May 1, 1888, Tesla was granted basic patents on his motor and the associate method of transmitting power by polyphase currents. Two weeks later, he delivered his classic lecture, "A New System of Alternate Current Motors and Transformers," before the American Institute of Electrical Engineers in New York.[4]

Westinghouse saw in Tesla's theories the certainty of a power coup. He offered thirty-two-year-old Tesla a million dollars for his patents. Tesla, who had been through the heights and depths, had no hesitancy in accepting. Tesla added the stipulation, however, that the payment of a dollar per horsepower royalty be written into the contract. Here were two plungers putting all their chips on the table for the biggest power gamble in history. The million dollars would not be Tesla's alone, for there were Mr. Brown and a few other helpers who would share with him. But even a half million made him a wealthy man with the means of following his chosen and expensive vocation.

There were some complications in the deal with Westinghouse in which the patents had to be broken up and reckoned individually at $25,000 per patent, but by July, 1888, technicalities were resolved and the battle lines were drawn between competitors.

Westinghouse arranged for Tesla to work as consultant for a year in his Pittsburgh laboratory. Tesla's financial worries were over and he was relieved to place the details in the practical and capable hands of Westinghouse.

In the Pittsburgh factory, there arose some friction between the engineers and the young inventor. Tesla was adamant in his choice of sixty cycles as standard frequency for the alternating current motor as opposed to the 133 cycles which had been the accepted standard of the engineers. Problems were beginning to enter into the single-phase operation. Artifices had to be incorporated to achieve some of the characteristics of the two-phase current. In his mind was the embryo idea of an improved polyphase system, but he was not working with the commercial urgency of the young Westinghouse engineers. Tesla could not brook opposition from the young Pittsburgh novices. Finally he left the Pittsburgh plant, but he later relished the news that the engineers had been forced to return to his sixty cycle idea, about which he had been so emphatic. Later, he declared that he had been of no help at the factory and that he could not work unless he was absolutely free.[5]

The kindling had already been laid for the electrical industrial conflagration. For some time there had been sparks flying among Edison, Westinghouse, and Thomson-Houston Company. Even in Europe, where there were many who held to the belief of the superiority of Edison's direct current, opinions were running rampant concerning the power battle that was exploding in America. It was common knowledge that:

> . . . leading . . . electrical scientists, Lord Kelvin and Werner von Siemans, like Franklin Pope and Elihu Thomson in America . . . warned against the use of alternating current systems. [Mistrust of the alternating current systems] arose from the supposition that, in the event of a transformer being short circuited or a part of the high tension system being accidentally grounded, the customer turning on a light might be instantly electrocuted. . . . Edison felt that the use of alternating current would bring down the whole electric light and power industry in ruins . . . that it would destroy the reputation for safety he had tried to build for his industry.[6]

The General Electric Company was founded, in part, on the war of the currents. In the battle, the Thomson-Houston and the Edison General Electric Company, the two biggest competitors of Westinghouse, merged to form the General Electric Company. They entered into psychological warfare tactics, with Edison commanding the firing line. He played upon the fears of a naive public who never suspected the invisible weapon being used upon them. They begged Edison, their defender, to protect them. Edison warned:

> Just as certain as death, Westinghouse will kill a customer within six months after he puts in a system of any size. He has got a new thing and it will require a great deal of experimenting to get it working practically. It will never be free from danger . . .
>
> None of his plans worry me in the least; only thing that disturbs me is that Westinghouse is a great man for flooding the country with agents and travelers. He is ubiquitous and will form numerous companies before we know anything about it.[7]

The press undertook to report the battle through editorials, informative articles, accusations, and rebuttals. The public read both sides and were helplessly caught in the struggle that was beyond their comprehension.

There was no doubt that Edison was swayed by passion and hatred for Westinghouse, whom he considered no better than a thief entering a private domain. Rage consumed this man of moods. All the hatred of which he was so capable extended to Tesla and the entire alternating current forces.[8]

The old ogre of capital punishment became a factor seized upon by the Edison followers to discredit the opposition. The reformers were already toying with the idea of electrocution as a more merciful way of meting out the death penalty. The state of New York appointed a commission to study the problem. This commission was headed by Harold P. Brown, an electrical expert. He was assisted by Dr. Carlos F. McDonald, medical superintendent of the Auburn Asylum for Insane Criminals; Dr. A. D. Rockwell, a noted electrician; and Dr. Edward Tatum of the University of Pennsylvania.[9]

Edison was opposed to capital punishment, but was of the opinion that it was a necessary evil. He also realized that if alternating current could be recommended for the operation, it might substantiate his constant claim that the "Westinghouse current" would have found its proper place and that it was far too deadly to be used in the homes of life-loving Americans. Here was a chance for a slightly below-the-belt blow at Westinghouse, so Edison very graciously offered the use of a building in the rear of his Orange laboratory for the research of the commission. If it seemed strange that Edison's direct current laboratory should be offered for experiment in alternating current, the only complaints came from the Westinghouse corner. Edison offered the assistance of his trained personnel in general and his brilliant staff representative, Arthur E. Kennelly, in particular.[10]

Edison was said to have rounded up the stray animals of the streets, some of which he bought for twenty-five cents, to be used in the experiments. Four dogs, a horse, and four calves were brought in to initiate the project. Charles Batchellor, who had introduced Tesla to Edison, was appointed executioner in the grim experiments. The first attempts were bungling

Thomas Edison and his original dynamo, Edison Works, Orange, New Jersey, ca. 1906. *H. C. White Co. photograph, Library of Congress Prints and Photographs Division Washington, D.C., (LC-USZ62-93698).*

and uncertain and dangerous, and Batchellor found himself badly jarred by one faulty test. Since alternating current was to be used as the death-dealing agent, and since the gigantic dynamos in use in the Edison laboratory furnished only direct, or continuous current, it was necessary for Batchellor to attach an alternating device to an electric generator used for street lighting and to develop a current of 1,200 volts.

In the laboratory, two wires were used—one attached to a sheet of tin and the other placed in a pan of water. A mongrel dog was brought in and it was expected that the dog, when placed on the tin plate, would reach to drink the water and in so doing would complete the circuit. According to the old maxim, horses cannot be made to drink, but that is not true of little dogs. When the dog backed off from the tin and refused to drink—it was forced. In due time the gruesome tests were ended to the complete satisfaction of all concerned at the Orange laboratory.

Westinghouse was far from satisfied with the experiments. With electrocution a certainty in the new future, he knew how tightly the propaganda net was being drawn. The State of New York completed preparations for the first use of the electric chair. A convicted murderer, William Kemmler, was selected as the first victim. His attorney, Bourke Cochran, fought desperately to save his client from this new horror. It was rumored that Westinghouse was fighting the experiment with his might and his money. Edison was called into court to give his opinion on the use of electric current to cause death. When queried, his replies were said to be vague, causing the attorney to demand:

> "You are testifying from belief, aren't you Mr. Edison? Not from knowledge?"
>
> "From belief," replied Edison dryly. "I have never killed anybody." Later Edison clarified his position by making plain his insistence on low pressure applied to densely populated districts—. . . high pressure wherever conditions are such that by no possible accident could that pressure get into the houses of the customers.[11]

Westinghouse could only fight with his purse to keep his "offspring of faith and courage from being turned into hangman's uses."[12]

The press smelled the smoke of a good story and fanned the flames higher. In the November, 1889, issue of the *North American Review,* Harold P. Brown, head of the commission, who had been proclaimed an authority on electrocution, reported the findings of the commission in an article titled, "The New Instrument of Execution."

He began by explaining to his readers the three classes of current—continuous (direct), pulsating-intermittent, and alternating. He stated that investigation had shown that the first two could be made safe and harmless to the general public, but that the third, or alternating current, the type that was being urged by Westinghouse, was by its very nature hopelessly deadly! He recalled that the previous year he had ventured such conclusions, and Westinghouse had fought back with a claim that his current was "safe."

The article indicated that Mr. Brown had been won over completely by Edison. He was loud in the praise of Edison's ability and generosity in placing his laboratory and staff at the service of the commission. Said Mr. Brown, "His enemies have never forgiven him for this great kindness, which permitted the proof that alternating current could produce instant and painless death at a very low pressure."

Mr. Brown defended his report further by saying:

> I felt that this use of the "safe and harmless current" could educate the public to handle it with caution and thus save many lives.[13]

Brown then proceeded to give a graphic description of all the details of electrocution, from the instant when a human victim's hands and feet were wet with potash solution up to the moment when the switch was closed and a man was actually beaten to a sudden death by the powerful contractions of his own muscles.

"Thus," said Mr. Brown, "the majesty of the law has been vindicated with no physical pain."

He concluded the article with a damning statement aimed directly at the eye-whites of the Westinghouse warriors, saying:

> Such is electrical execution, and yet, strenuous attempts have been made to befog the public mind in order to prevent the use of alternating current for the death penalty lest the public learn of its deadly nature and demand that the legislature banish it from streets and buildings, thus ending the terrible, needless slaughter of unoffending men.[14]

Many "unoffending men," reading these words, agreed with Brown that the legislature should do something!

Such a barrage from the Edison lines drew return fire. George Westinghouse, Jr. fought back in the December issue of the *North American Review.* His article was titled, "A Reply to Mr. Edison." Apparently he considered the November article, which had been signed by Harold P. Brown, to be his in name only. So Westinghouse directed his reply to the power behind the pen. Westinghouse began:

> Electric lighting, unlike other industries protected by patents, has been followed with keen interest by the public at large, but among the more intimately interested parties, the struggle for control of the electric light and power business has never been exceeded in bitterness by any of the historical commercial controversies. Thousands have pecuniary interest.[15]

He pointed out Edison's penchant for cut-throat competition, quoting a *New York Herald Tribune* report in which Edison had been credited with saying, "I don't care so much for a fortune, . . . as I do for getting ahead of the other fellow!" Westinghouse quoted Edison as saying, "My personal desire would be to prohibit entirely the use of alternating current." He called attention to the inconsistency of such concern for safety on the part of his adversary when the Edison Company made it common practice to use uninsulated overhead wires carrying 220 volts. He assured the public that the roasting of a large piece of beef was possible with Edison current of less than 100 volts in two minutes. He quoted the admission of death-potential from direct current using the words of Edison's staff man, A. E. Kennelly, in a previous statement.

He accused the commission of misrepresentation because they had not employed the alternating current that was used for commerce, but said that it was Edison's direct current, made alternating by a pole changer, thus producing an effect incomparably more dangerous than the true alternating current. He explained that the high voltage street wires had been tapped.

He concluded:

> Mr. Edison has always said, that in the long run, every system will fail which does not for domestic service use low pressure current. This is exactly what alternating current supplies.[16]

Edison called, "Foul." He repeated, "The trouble with "Westinghouse is that he will spread his ideas all over!" Edison was right.

The public were more confused than ever. What were they to believe? Alternating current chosen for the lethal element in the electric chair made a very convincing argument. They begged the press to intercede again with Edison to save them. Edison had scored.

Even national economic factors seemed to ally themselves on the side of the promoters of direct current and money power. Little companies merged, preferring to be swallowed rather than to face competitive ruin alone. Financial conditions pushed Westinghouse into a bind. Every corner had to be cut for his company to survive. The Westinghouse Corporation looked hungrily at the dollar per horsepower royalty in Nikola Tesla's contract. It is believed that the company decided that Tesla would have to help them in their extremity. George Westinghouse, Jr. was chosen to approach Tesla with the ultimatum of surrender of the horsepower royalties, or to face defeat with the company. The two men discussed the situation at length. No matter how they tried to solve the problem, the answer came out the same. Tesla hated to give up the income assured in his contract—yet what good was a royalty right with a financially ruined organization? There had been a time when he had experienced the depths of poverty and despair. In this dark moment, George Westinghouse, Jr. had been the only one to hold out a hand to him, and the only one willing to pay him an Alger-like fortune for his patents.

Tesla reached for his contract. It was a moment of hard decision. He made his answer without whining—and as a gentleman. He tore the contract into bits and dropped them in the waste basket. It has been estimated that in so doing he gave up a potential return of twelve million dollars in his lifetime.[17]

Reliving the incident in a speech at the Biltmore in 1932, he explained:

> George Westinghouse was in my opinion, the only man on the globe who could take my alternating current system under the circumstances then existing and win the battle against prejudice and money power. He was a pioneer of imposing stature and one of the world's noblemen.[18]

There is a story that Tesla also assisted Westinghouse in combatting the psychology of fear spread by the Edison forces. This he did by means of one of his dramatic exhibitions to which the press was invited. The refutation of the danger of alternating current was staged in Tesla's laboratory. A few reporters feared that the inventor might lose his life that night, since his experiment was to entail lighting of lamps by his high frequency current, which he intended to pass through his body.

Tesla must have looked like a suave magician as he appeared on stage that night, before a hushed and fear-excited audience. He was dressed in white tie and tails and tall silk hat—his customary dress for such demonstrations. It was a gesture of honor and respect to the electrical force which dominated his life. As he successfully performed the weird experiments, lighting globes and tubes with the current flowing through his body, he reassured the public of the safety of his current. He proved to them that it was not high voltage which destroyed flesh, but the high amperage or current density.[19]

Often compared to a magician's performance, Nikola Tesla amazed the public with his demonstrations of alternating current. His intent, to educate the audience on the remarkable safety and potential of electricity, was often overwhelmed by the showmanship and entertaining nature of his performance. *Courtesy of Fran Kiester.*

Tesla was not only a figure in the industrial spotlight, but he was also in great demand at this time in the social circles. Society was completely won over to the suave, handsome young scientist. He was a gracious host and also a sparkling addition to any dinner party. Mothers of marriageable young daughters considered him seriously as a potential acquisition to their family trees. Although the young foreigner could offer no title, his polish, his Continental manners, and his half-million dollars weighed heavily in the balance.

The Waldorf-Astoria Hotel, New York City, ca. 1901. *Library of Congress Prints and Photographs Division Washington, D.C., (LC-USZ62-77182).*

Tesla often gave dinner parties at Delmonico's and the Waldorf Astoria which were characterized by good food and brilliant conversation. The elite could be assured of a preference on his guest list. His less formal evenings at home were revolutionary in entertainment and unique in food, which he enjoyed preparing himself. In his early years he had been a gourmet, dining on saddle of lamb, baby lamb chops, roast squab with nut stuffing, and roast duck smothered with celery stalks. All of these he insisted on personally supervising in the kitchen. Of his favorite roast duck, he ate only the choice morsels on either side of the breast bone.

Although he ordered lavish meals for his friends, entertaining them graciously, he was gradually withdrawing from such self-indulgence. He ate more and more sparingly, particularly of meats. His love of coffee presented his greatest problem of abstinence. He had long ago decided that it was not beneficial to him, but he found it difficult to forego the aroma. For ten years he would order a pot of his favorite coffee and would have it poured, but would not drink it. He gave up smoking, and drank whiskey temperately. He believed the whiskey beneficial, but he felt no need for it as a crutch.

A Tesla dinner was, for his guests, a command performance. The guest list was a part duplication of Ward McAllister's "400." The meticulous host would sample all food in monarchal fashion, to assure himself that it met the perfection he demanded for his guests. Almost always, some sauce or wine failed the test and was sent back to the kitchen.

No less dramatic was the after-dinner, amazing entertainment when the guests repaired to his laboratory to witness the latest in his electrical science shows. The diabolic devices from which issued unearthly lights, sulphurous fumes, glowing colors, and frightening noises lent an aura of mystery and fear that was mitigated only by the confidence in the star performer. When thousands of volts passed through his body to melt wires or light lamps, this was magic, black magic, and only a few of his guests were interested in the scientific explanation. For Tesla, phenomena always demanded scientific explanation. There were no mysteries for him that could not be reduced to formulas. He saw miracles not as something against Nature's law, but as something that had managed to operate by some unknown law of nature. Tesla often remarked, "Look for the law behind the phenomenon."

This was a pleasant period of Tesla's life. He enjoyed his friends. Most of all he found it good to be back in his own laboratory quarters on lower Fifth Avenue. It was a relief to be working alone again, away from the friction and confusion of the Pittsburgh factory. An inventor could not be constantly harassed by argument and conflict with young men who were not his equal in his field. He was convinced that his world of discovery required absolute freedom. There were many young women now to make life pleasant. Yet, he dared not yield any place in his life to any woman. Women were too distracting. His was an uncharted path which he must walk entirely alone. As Thoreau had once written, not every man could conform his pace to others. Tesla was convinced that he must walk to the "rhythm of his own drums."

Notes

1. John J. O'Neill, *Prodigal Genius*, (New York, N.Y.: Ives Washburn, Inc., 1944), 65-66.

2. Francis E. Leupp, *George Westinghouse*, (Boston, Mass.: Little, Brown & Co., 1918), 140-141.

3. Merrick Jackson, *George Westinghouse*, (New York, N.Y.: Mercer Publishing Co., 1956), 7.

4. Kenneth M. Swezey, "Nikola Tesla, Pathfinder of the Electrical Age," *Electrical Engineer*, September 1956, 2.

5. O'Neill, *Prodigal Genius*, 76-77.

6. Matthew Josephson, *Thomas Alva Edison*, (New York, N.Y.: McGraw Hill Book Co., 1959), 345-346.

7. Thomas A. Edison, "Dangers of Electric Lighting," *North American Review*, November 1889, 343-345.

8. Josephson, *Thomas Alva Edison*, 345-346.

9. W. K. L. Dickson and Antonia Dickson, *The Life and Inventions of Thomas Alva Edison*, (New York, N.Y.: Thomas Y. Crowell Co., 1894), copyright by *Cassiers Magazine*, 326-331.

10. Ibid.

11. William Adams Simonds, *Edison, His Life, His Work, His Genius*, (New York, N.Y.: Bobbs Merrill Publishing Co., 1934), 154.

12. Leupp, *George Westinghouse*, 154.

13. Harold P. Brown, "The New Instrument of Execution," *North American Review*, November 1889, 586.

14. Ibid.

15. George Westinghouse, "Reply To Mr. Edison," *North American Review*, December 1889, 653.

16. Ibid., 655.

17. O'Neill, *Prodigal Genius*, 83.

18. Ibid.

19. Helen B. Walters, *Nikola Tesla*, (New York, N.Y.: Thomas Y. Crowell Co., 1961), 99.

A coil placed in a field near Tesla's Colorado Springs laboratory received electrical energy transmitted through the ground, sufficient to light a lamp, demonstrating the possibility of the wireless transmission of energy. *Courtesy Nikola Tesla Museum, Belgrade, Serbia, (006-02-148).*

Home of Lucien Lucius Nunn, Telluride, Colorado. *Photograph by Wanetta W. Draper, courtesy of Fran Kiester.*

Telluride, Colorado's Contribution to Electrical World History

Eastern industrialists were far from convinced by Tesla's showmanship nor did they have much confidence in Westinghouse's claims of the practicality of alternating current. Yet there was some imminent solution toward which the electrical scientists of the world were groping. The fever of the power race was felt in Germany and Australia, as well as in America. Strangely enough, the *initial* success was not to be found in any of the large centers of the world, but was to be worked out in an obscure mining area in the high country of western Colorado.

Westinghouse had reached a stalemate in the east and he found little opportunity for proving his ideas. Little did he suspect that a *world's first* solution would be brought about through the efforts of a five-foot-one human dynamo of masculine energy, Lucien Lucius Nunn. This young Colorado lawyer offered to Westinghouse the perfect testing ground for his disputed theories. Colorado was far enough away to be unimportant to the forces of the opposition. Western pioneers were congenital gamblers, and these particular gamblers were willing to put all their chips on the turn of the cards. Nunn and his associates were compelled by the unaccountable optimism of the pioneer—and by a dire necessity.

Nunn had pitched his tent in the brawling little mining town of Telluride in 1881. Legend maintains that he started his fortune by renting his zinc-lined bathtub at fifty cents a dip to miners headed for the "Pick and Gad" on a Saturday night; legend also declares that he became a lawyer when he was hit over the head by a law book in a barroom brawl.[1] Actually, neither of these is quite accurate, since he had spent some time in the study of law at the Universities of Leipzig and Goettingen.[2] His famous bathtub was the first in Telluride, but it was planned for his own use.[3] However, he did operate a bathhouse on the side, with a charge of fifty cents per bath.[4]

The climate of Colorado was a favorable one for Nunn because of his frail health. He had plunged into the restaurant business in Leadville in 1880, where he served the choicest food for twenty-five cents a meal. Later on, he branched out into a more pretentious establishment, giving it the fancy name of Pacific Grotto, and doubled the price of the meals. The exorbitant price sent the customers back to the sow-belly and beans of the other local eating houses and to Nunn's own more economical Leadville House. His extravagant gesture was doomed. He moved from Leadville to Durango, where he nailed up his shingle as a lawyer and operated a restaurant on the side.[5] He did a great deal of carpentry, but it was strenuous work for his slight frame. His light auburn hair denoted a fiery temperament not in keeping with his slight physique. His pale blue eyes, light skin, and small stature set him apart in a country of massive mountains and rawboned, tough-skinned giants. With the increased mining excitement around Telluride, he traveled on foot across the mountains. He soon got out his hammer and saw again and nailed together furniture, took on a job of roof shingling, and built and furnished a lodging house. All this on an economical diet comprised chiefly of oatmeal. When he had accumulated sufficient funds, he opened his law office of Nunn & Kinney.[6] As he began to prosper, he built a home for himself which still stands on the corner east of the Telluride High School.[7]

It was the crisis in mining in the Telluride district which plunged Nunn into speculation as to the solution of the power problem at the mines. The Gold King and other high altitude mines had been operating on steam power, but the problem of fuel was rapidly driving them into bankruptcy. The mine owners turned to Nunn for legal advice and for rescue from their problem.

Because the mines were situated above timberline and had depleted the nearby sources of fuel, they were forced to go farther and farther for their supply. Coal was selling for around fifty dollars a ton and had to be packed in on burros. The surface ores were becoming exhausted, and the increased depth of the tunnels raised the cost of operation.

Nunn became a manager for the eastern interests in the Gold King mine in 1888. His partners were James Campbell, one of the wealthiest men in St. Louis, and Benjamin Butler, a criminal lawyer and politician, a former governor of Massachusetts.

It was obvious to Nunn that the crux of the problem was cheaper power. He had long been interested in the pioneer experiments of George Westinghouse. Although Nunn himself was not an engineer, he knew where to put his finger on a good one. In 1890, he sent for his brother, Paul N. Nunn (later chief engineer for Ontario Power Company, Niagara Falls).[8]

The partners realized that the source of cheap power for all the mines in the area lay in the valley below the mine.

> The south fork of the San Miguel and its descent from Ophir, a mining camp near Telluride, falls five hundred feet in a distance of less than a mile. The river is fed almost entirely from springs in the Ophir basin, and even at the season of lowest supply can furnish about 2000 h.p. Here was available power enough to operate every mine in the vicinity.[9]

The engineers looked over the territory and a contract was signed with Westinghouse. At that time

> All existing motors and other electrical equipment [for this purpose] were designed for direct current operation, which meant that everything to do with alternating current development had to be adapted, improvised, or invented. Easterners were snickering at the fantastic scheme and prominent engineers predicted it would be a miserable failure.[10]

Above the beginning of the rapid descent of the stream a dam was built in an almost perfect spot and a plant was erected at Ames, in the valley. A wooden shack was constructed and a six-foot Pelton water wheel furnished power to a 100 h.p. generator, 133 cycle, single-phase, 3,000 volts. It was belted to the wheel. The wheel operated under a 310 foot head, and a motor identical to the generator was installed at the mill, 2.6 miles away as a crow flies. Both the generator and the motor were Westinghouse alternators, the largest then made, designed from Tesla's patents, and bearing his name.

The Electrical World jumped the gun and published an account of what was going on in Colorado in their issue of March 21, 1891. This article also gave two pictures of the motor and generator of the Telluride mining plant and described them as Tesla motors. It listed the conducting circuit as a number three B & S gauge bare copper wire, erected on stout poles at sufficient distance above the ground to prevent accidents. It estimated the loss in transmission would be barely five percent. *The Electrical World* declared:

> this Telluride plant promises to be, when it shall finally be put into operation, a few weeks hence, one of the most interesting mining plants in the world from the nature of its location, which made electrical power a necessity, and especially from the daring way in which the difficulty of using very high potential has been met by employing a synchronizing motor . . . the plant is now rapidly nearing completion.[11]

News of the experiment in Colorado astounded industrialists in the east, but did not convince them. George Westinghouse, Jr., did not underestimate the importance of this project to him and dug down in his pockets for $25,000 to augment the experiment being carried out by trial and error.[12]

The switchboard was beautifully made of oak 2x4s and 1x3s and the oak was heavily paraffined. In the center of the board was a circuit breaker, one that was hardly automatic in the modern sense of the word. When a circuit was to be broken with the power on, the operator took hold of the plug and pulled it out. As Paul Nunn recounted later: "You can imagine, I fancy, how you would feel in pulling out a plug like that, running back eight feet and finding the arc still holding."[13] If that didn't work, a cap or hat was used to fan it out.

Electric storms were a constant hazard and after generator coils had been replaced a few times, measures were taken to avoid breakdown. The engineers placed oak 4x4 pieces as insulating platforms under the generator and the motor. These oak pieces they first boiled in paraffin, although they had been told that it was impossible to paraffin oak. By running paraffin at the boiling point of water for twenty-four hours, they drove out the vapor from the sticks and the vapors were condensible. The sticks were kept under the surface of the paraffin until they cooled. A vacuum formed within the fiber of the wood and soaked the paraffin into it for the depth of several inches. Still there was lightning. Some form of lightning arrester had to be devised. Sometimes the strokes would come four or five a minute. Nunn said that these might be called "atmospheric disturbances," but that they didn't know the difference.

One form of lightning arrester which they devised was a marble box with two holes in it, through which pickaxes were dropped, then the points of the pick equipped with pieces of arc-lamp carbon. The explosion from the lightning discharge would kick out the pickaxe and break the arc. This was probably effective even when they weren't aware of it.[14]

Once the plant was in actual operation in 1891 it ran continuously, smoothly, and steadily for thirty days without stop after its initial start. The plant operators were admittedly afraid to shut it down for fear they would be unable to start it again.[15]

As Nunn later reported:

> The first instruments resembled an assayer's balance. One end was a bundle of loose iron wires. On the other end the counterweight. The telltale swung back and forth on the scale. Theoretically, it was correct, but if anything happened to tip the instrument or to shake the switchboard, the calibration was all off. We didn't know much about the calibration in those days. We set up a thing we thought was right and took for gospel truth whatever it read. A watt meter? We had none. We did not know what watts were, and we would not have known what to do with one. Transformers up to this time had been rated in "lights."[16]

There was not only the problem of devising equipment but of manning the station with workers who knew how to handle it. They could not hope for graduate engineers since there was only one college in the country (Ohio State) at that time which had a well-grounded course in electrical engineering. The difficulty was to find young men pliable enough to discard old theories for uncertainties, to accept the challenge of pioneering in an industrial upheaval in which they were to be the front line.

An educational plan became an integral part of the operation of the company. Students were also employees. At first the school was housed in Telluride, next to Nunn's home. Known as the Telluride Institute, the building still stands. Students were recruited from Cornell to mingle with the boys from the frontier. From the experiment grew the Telluride Association, a unique idea in education. The first courses taught something of machinery, shop work in metal and wood, wiring, insulating, and repairing. The boys were furnished a technical library which included electrical papers and a conveniently fitted testing room. Each was given a laboratory course in alternating current theory. Pay was thirty dollars a month with room and board. So was born the first systematic idea of on-the-job training by a corporation.[17]

Training was not limited to theory or pencil work. The boys put on their overalls, climbed the poles, strung the wires, and did the repair work. It was a course of trial and error, of learning by doing. It was a rugged two-year experience for the "pin heads," as they were called, but not without its lighter moments. The boys noticed that the frequent electrical discharges which were the bane of their lives were not limited to human inconvenience. At one time when a boisterous old bull came down to graze on the valley grass, he headed for the river to drink, near the place where the boys had placed a lightning arrester. At that moment there came a thunderstorm. Paul N. Nunn afterward related the effect on the bull's abdominal muscles or his diastolic action saying "the curve was very straight."[18]

The students of 1891–92 decided one day to extend their electrical experiments to the problem of eliminating the snakes from the powerhouse. There had been difficulty with these disconcerting reptiles, which had a habit of crawling in during the night. Closing the door had not solved the problem, since the snakes immediately found a way through the wall where the operating shaft extended to the water wheel. So one bright-eyed lad suggested a solution of electrocution by placing two metal plates over which the snakes must crawl before they got into the building. The theory was correct. The snake was electrocuted. Unfortunately, the side effect was a short circuit which shut down the plant.[19]

When the operation moved into Provo Canyon, Utah, the school was taken there and later formally incorporated as the Telluride Association. The educational program was extended, over a period of time, to include such subjects as history, English, German, algebra, and physics.[20] Six years after forming the Telluride Association, L. L. Nunn endowed the Deep Springs School in California.

At Olmstead Power Station, at the mouth of Provo Canyon, Utah, the most imposing structure was the Telluride Institute, completed in 1904. The building contained classrooms, library, laboratories, and living quarters. The work was carried on here for seven years until a need arose for the advanced students to continue study at a university. On the Cornell campus at Ithaca, New York, a building patterned after the Telluride House was erected in 1909, designed to be an independent branch of the university.[21] At the present time it is a self-governing association, with a common purpose of education and a responsible attitude toward the duties of profession and citizenship—not based on a common professional interest. Each year thirty men studying at Cornell are awarded room and board scholarships to live at Telluride House.[22]

Thus Nunn's dreams and plans for an educational plant are still being carried out. He was a man of deep convictions, a strict moral code, and a firm belief that learning should be chiefly concerned with the building of a man, in which education was a means but not a goal. The Institute no longer restricts itself to engineering students, but it requires a high potential of a man, in keeping with the ideas upon which it was founded.

The Ames plant, after seventy-one years, is now part of a giant electrical system. Incorporated first in 1891 as the San

Miguel Consolidated Mining Company, it became Telluride Power Transmission Company, and then the Telluride Power Company. On May 1, 1913, it was absorbed into the Western Colorado Power Company and in August, 1962, became a subsidiary of the Utah Power and Light Company.[23]

> Today, the Western Colorado Power Company has grown into an interconnected power system of eight power plants with over 600 miles of transmission lines serving an isolated empire in southwestern Colorado of 50,000 people. The output today is around 50,000 h.p. of energy. All connected with the original Ames project became rich and their names have gone down in electrical history as true pioneers. Little did they realize that their efforts to harness that small stream of water near Telluride would eventually lead to the great electrical industry as we know it today.[24]

The most recent figures exceed the 50,000 subscribers listed in 1961.[25]

The rotor from the first machinery at the Ames plant lies rusting on the ground, stripped of copper. The laminated sections have sunk into the ground and grass grows in the dirt which has accumulated on it. Just recently (1963) history-conscious residents of nearby Telluride have made efforts to establish a museum with replicas of the early-day equipment of the Ames plant. In the valley, near the original power plant, live two families.

So ends the story of a world's first which had its birth in Telluride, Colorado. But it was only first by a margin of a few short months. Engineers in Germany were working feverishly, with Tesla's patents, to complete a power line from Lauffen-on-the-Neckar to Frankfort-on-the-Main under a heavy penalty to complete the project by August 15, 1891—a distance of a hundred and twelve miles.[26] However, they were unable to meet the deadline and the preliminary trial was not run until August 26, 1891.[27] It was important enough news to rate a cablegram from the Frankfort correspondent of the *Electrical Engineer*, and they stopped the presses to run the news in the issue of that date.

The Edison Company, in a flurried attempt to push into the booming Colorado development, hastened to complete a project of its own in that same area. It was March, 1892, before they were able to complete and operate a power station located four miles from the Virginius Mine, near the summit of Mount Sneffels—an altitude of nearly 12,700 feet. This station, near Ouray, generated 245 h.p. secured from Red Canyon Creek by direct current.[28]

Although Australia was playing with the idea of alternating current at this time, long distance high voltage power-transmission was not involved. The Thomson-Houston Company was spreading into that field and Westinghouse had gone into Newcastle in that area only a few months before.

Actually, the race for efficient electrical installation by various companies and methods was world-wide and swift. There are some authorities who place the date of the Portland, Oregon, project in 1890. However the *Electrical Engineer* of August 31, 1892, dispels that possibility, placing the date of the organization of the Portland Company as late as 1891. It is hardly plausible that it could have predated Telluride for completion. Telluride was also in construction in 1890 but its first date of operation is definitely June, 1891.[29] Italy and Germany were running neck to neck and Switzerland was also in the money in the power race. The *Electrical World* soon declared that the successes at Telluride and Portland indicated that the use of alternating current was fast passing from experimental investigation to practical engineering. Of all of these developments, the Frankfort-Lauffen held the most world-wide attention.

Our final estimates must place the time-lead with the Ames plant at Telluride, which began June 21, 1891, as the first power station in the world to transmit alternating current at high voltage for power purposes, for a long distance. This, of course, is a highly-restrictive statement.[30]

Half a decade later, Telluride was to be given credit for her contribution of experience in the decisions at Niagara. At that time, Lyndle Stetson, first vice president of the Cataract Construction Company, would write in *Cassier's* July issue, "Tivoli, Turin, Telluride, Genoa, Willamette, San Bernardino—

all tell that commercial success lies back of that brilliant experiment of Lauffen and Frankfort." And, of course, the Frankfort-Lauffen was a factor in the decisions for Niagara power development.

In the fever of the excitement at Telluride, any honor to Tesla was by-passed, although his patents were basic to all this progress. He shrugged off the ingratitude of the world, so intent was he on the invention at hand and the others spinning in his brain. Among the inventions that year of 1891 was the Tesla coil, designed to produce high voltages, "a device which remains today in one form or another in every radio and television set."[31]

The cut-throat competition evolving from the mad scramble for financial power made larceny of patents and ideas the accepted practice. A man who would not think of picking pockets had no qualms about picking brains. Elihu Thomson claimed that it was he who had interested Tesla in alternating current, although Tesla's interest had stemmed from the early school days at Graz. Thomson insisted that the Tesla coil was misnamed and that it involved infringement on the Thomson patents.[32]

Patent lawyers grew fat during this era. But this was not the first nor the last time that Tesla was to be accused and acquitted.

The highlight of the year for Tesla was the granting of his United States citizenship. For five years, since his application, he had waited for this time. Now he would no longer begin his patent application, "I . . . a subject of the Emperor of Austria-Hungary." Now he could write, "I, Nikola Tesla, a citizen of the United States." He put his precious citizenship papers in his safe, while he stowed away his orders, diplomas, degrees, and other distinctions in his old trunks.[33] Thus with his American citizenship, he severed a cord with old world ties.

Rotor from the original armature at the Ames Power plant, a part of Westinghouse machinery using Tesla system. *Photograph by Wanetta W. Draper, courtesy of Fran Kiester.*

Notes

1. Perry Eberhart, *A Guide to the Ghost Towns and Mining Camps of Colorado*, (Denver, Colo.: Sage Books, 1959), 323, 327.

2. Stephen A. Bailey, *L. L. Nunn, A Memoir*, (Ithaca, N.Y.: Cayuga Press, Printed for Telluride Association, 1933), 41, 51.

3. Ibid., 51.

4. *The Telluride Times*, May 17, 1963, 2.

5. Bailey, *L. L. Nunn, A Memoir*, 46-48.

6. Ibid.

7. Telluride Chamber of Commerce, Telluride, Colorado, pamphlet.

8. *75th Anniversary Issue of Electrical West*, August 1962, Vol. 129, No. 2, 299.

9. "A Unique Mining Plant," in *The Electrical World*, March 21, 1891, 223.

10. *75th Anniversary Issue of Electrical West*, 299.

11. *The Electrical World*.

12. Mrs. M. L. Cummins, *Pioneers of the San Juan*, Vol IV., (Denver, Colo.: Big Mountain Press, 1961), "An Industry and An Institution of Higher Learning Are Born At Ames, Colorado, 1891." Printed for Sarah Platt Decker Chapter, NSDAR, Durango, Colo.

13. *75th Anniversary Issue of Electrical West*, 299.

14. Paul N. Nunn, "We Did Not Know What Watts Were," *General Electric Review*, September 1956, 46.

15. Ibid.

16. Cummins, *Pioneers of the San Juan*.

17. Bailey, *L. L. Nunn, A Memoir*, 71-72.

18. Nunn, "We Did Not Know," 46.

19. *75th Anniversary Issue of Electrical West*.

20. Cornell Branch of Telluride Association, pamphlet, 7.

21. Bailey, *L. L. Nunn, A Memoir*, 98.

22. Cornell, 2.

23. Cummins, *Pioneers of the San Juan*, plus correspondence with C. P. Schools and Art Wyatt.

24. Ibid.

25. Ibid.

26. *Electrical Engineer*, Vol. XI, June 10, 1891, 648.

27. Ibid., Vol. XII, August 26, 1891, 263.

28. *Engineering News*, March 12, 1892, 239-240.

29. Monk Tyson, "The Gold King Was First," Empire Section, *Denver Post*, October 7, 1962, 10.

30. The mention of Great Barrington, and its contribution as an initial experiment in alternating current has been made in a previous chapter. It is not included with the 1890 decade of alternating current accomplishment, since the progressive step of that era was the factor of distance transmission. Great Barrington celebrated its alternating current success, March 23, 1886, when the people gathered in the town square to watch the lights go on in thirteen stores, two hotels, two doctors' offices, the telephone exchange, and the post office. Some industry was involved but it was all within the close area of the power source. Nevertheless, this town is an important shrine in electrical history.

31. Kenneth M. Swezey, "Nikola Tesla, Pathfinder of the Electrical Age," *Electrical Engineering*, September 1956, 3.

32. David O. Woodbury, *Beloved Scientist*, (New York and London: Whittlesey House, 1944), 180.

33. Nikola Tesla, "My Inventions," *Electrical Experimenter*, June 1919, 178.

The original Telluride Institute, Telluride, Colorado. *Photograph by Wanetta W. Draper, courtesy of Fran Kiester.*

Columbian Exposition—the grand court at night—electrical illumination of MacMonnies's fountain and the administration building. *Charles Graham drawing, from* Harper's weekly, *June 24, 1893, Library of Congress Prints and Photographs Division Washington, D.C., (LC-USZ62-100741).*

Lighting the Columbian Exposition

The year of 1892 found Tesla abroad. He was in demand as a lecturer before the best scientific societies of Europe. French papers were full of praise for Tesla's brilliant achievements, and they declared that no man in that age had won such a universal scientific reputation in a single stride.[1] His reception at the hands of electrical engineers had become, like his inventions, a part of electrical history. Newspapers maintained that the honors conferred on him were such as to make Americans very proud of one who had chosen America for his home.[2]

Part of his tour was planned so that he might visit his mother and sisters and his home country. Tesla had always closely identified himself with his mother. The two seemed to have a strange telepathic relationship, so Tesla became very concerned one evening when he felt a premonition in which he seemed to know that his mother was dying. A telegram came shortly after this and confirmed his fears. Taking a train to Gospic, he reached there only a few hours before her death. The shock of the loss of the only one who had really been close to him, compounded with the physical strain of his lecture tour, and his inability to sleep, brought on a serious illness which delayed his return to the United States until August.

The death of his mother became a turning point in his life. During his convalescence, he took stock of his abilities and the uses to which he had put them and he was dissatisfied with what he found. He realized clearly that his lectures and social life had absorbed too much time. He had been hungry for human contact and acceptance, but in satisfying this hunger he had dissipated his abilities. He knew the emptiness of acclaim, pleasant as it had been. More and more, he had come to the conclusion that an inventor should live in a world of his own. Work could be his only constant companion.

Tesla returned to America to find that much had been going on in the engineering field. Plans for completing the details of the great Columbian Exposition in Chicago had really stirred up a hornet's nest. The exposition, to commemorate the discovery of America, had been postponed a year because of the Presidential campaign. This was the first World's Fair that had ever had the opportunity to be lighted electrically, and the big companies had seen the advertising possibilities of electrical displays that could be dramatically lighted at night against the background of architectural beauty, landscape gardening, and water effects. Bids on the electrical power plants for the fair had been called for to be submitted before April. The first request for bids stated that the generators were to be of the direct current type.[3]

On March 12, further news announcements concerning bids indicated that alternating current might be considered. The *Engineering News* had published an article which contained a paragraph which began, "If alternators are to be used" and concluded with safety measures and requirements which would be demanded in the event that alternators should enter into plans.[4]

It was commonly believed that only two companies were competent to undertake a contract for such an enterprise—General Electric and Westinghouse. To the amazement of the commission, a dark horse entered the race with a lowest bid. The surprise bidder was Charles Locksteadt, president of the South Side Machine and Metal Works of Chicago.

"Who is Charles Locksteadt?" asked officials. "How did he manage to get into the starting line to crowd the tape?" There was an uncertainty that this man could fulfill such an important contract. The commission decided to call for new bids.

Locksteadt went to Westinghouse and asked him for support, with the result that Westinghouse decided to lower the Locksteadt bid. He determined to make sure of securing the contract by reducing the (already too low) bid of $5.49 per light to even less. No company could meet this bid legitimately and come out even. This was, on the surface, another instance of Westinghouse daring and recklessness. Actually, this far sighted genius was demonstrating plain horse sense by taking the long view. He could do it for nothing and come out ahead in prestige and world advertising.

On May 26, 1892, the *Engineering News* announced:

> The contract for furnishing the incandescent electric lighting plant has been awarded to the Westinghouse Electric and Manufacturing Company of Pittsburgh, Pa., at $399,000 for 96,620 lamps at $4.32 a lamp, using the alternating current system. The original bid of Edison, Thomson-Houston and other electrical companies was $18.50 per lamp or $1,713,507. An outside company bid $5.45. A similar bid of $38.50 per lamp was made for the arc lighting plant. Here again, defeated by outside bids, but eventually contracts at $20 per lamp were let to the Thomson-Houston and allied companies and the Western Electric Company for one thousand lamps. Foreign companies had been invited to furnish the balance at the same price and Mr. Sargent, the electrical engineer made a trip to Europe to confer with those companies.[5]

The story of the bitterly contested fight to light the World's Fair thus was concluded, and on July 21 the *Engineering News* reported:

> . . . the contract for the electric light plant was awarded to Westinghouse and satisfactory bonds have been given and the contract closed. Generators to furnish the electricity for 93,000 incandescent lamps will be the largest in the world. One will operate 20,000 lamps and most of the others, 10,000 lamps. Ordinary generators supply 1,200 lamps.

Walking away with the contract for the incandescent lighting, signalled only the beginning of the problems for Westinghouse, however. Edison held all the patent rights in existence on the only incandescent, all-glass globe, and there would be no cooperation from that quarter.[6]

The challenge was to invent an entirely new lamp in eleven months and eight days. Westinghouse would have to avoid infringements on existing patents on the one-piece style lamp owned by Edison. He set up a glass factory in Alleghany and produced a two-piece stopper lamp. In this lamp, a ground glass stopper containing an iron filament fitted into the bulb like a cork.[7]

The Edison forces were eating crow and the diet was not to their liking. The small sop thrown to them in the arc-lighting contracts at prices below their expectations was something, but not enough. Neither were they satisfied with the Westinghouse plan to by-pass the Edison incandescent lamp. Both sides were at each other's throat. Westinghouse happened to overhear a conversation on the train one day, between two of Edison's attorneys. He learned that one of them planned to be in Pittsburgh the following day. Since he had seen the man just the day before and no mention had been made of this, Westinghouse suspected that something secretive was impending. He called his legal advisor to get in touch with Leonard Curtis, his patent attorney in Englewood. Curtis wired a colleague in Pittsburgh to look out for whatever was in the wind.

The Pittsburgh colleague "just happened" to be in court when the attorney for Edison appeared. When the judge arrived, it was revealed that the Edison companies had applied the day before for a restraining order to prevent the Westinghouse Electric & Manufacturing Company from selling or otherwise disposing of their electric lamps. They charged the Westinghouse Company with bad faith and resorting to subterfuge. Westinghouse was cleared of any attempts of evasion.[8]

These constant battles were fought with bitterness, collusion, and chicanery. It is doubtful if either side missed a trick. An anecdote of the day was that at one court session a document held by the Westinghouse attorney was turned over by a gullible judge to one of the Edison attorneys on the plea that he needed to examine the paper before the next session. The judge admonished him that he could do this only with the solemn promise that it would be returned in court the next morning. The next morning the attorney appeared but declared he was unable to find the important paper although he had searched diligently.

The judge countered, "How can you search diligently for something you obviously do not want to find?"[9]

Such was the state of the "Battle of the Currents" when Tesla returned to the United States.

By the time the fair opened, May 1, 1893, Westinghouse had 250,000 lamps ready to illuminate the grounds and his was one of the few large installations that was in place and complete. It included twelve dynamos, ten feet high, weighing about seventy-five tons each and constructed on the Tesla multiphase system.

The Statue of the Republic silhouetted by thousands of incandescent lamps electrified by Nikola Tesla's multiphase alternating current. The "White City" was destroyed by fire in January 1894. *From Kenneth M. Swezey Papers, Archives Center, National Museum of American History, Smithsonian Institution (AC0047-0000009).*

One of the most popular exhibits at the fair was Tesla's personal one in which he demonstrated a spinning egg made of metal. The metal egg was placed on top of a velvet-covered round platform. When the switch was closed, the egg stood on end and rotated rapidly. By this, Tesla tried to explain the principle of the rotating magnetic field, but to the wide-eyed crowd it was only a trick. Like the mob in the days of old Rome, they asked only for bread and circuses—and this was a circus. They did not try to understand Tesla's polyphase system.

Even more mystifying was the demonstration at the fair which had often thrilled Tesla's New York audiences. When one million volts of alternating current of high frequency passed through his body, it was an answer to Edison's accusation that alternating current was deadly.[10]

So, Tesla, the mystery magician at the fair, tried to explain to the curious throng the principles of a field in which the scientific world was still groping. He accepted an invitation to speak before the National Electric Light Association in St. Louis. His audience was so large that the meeting was moved to Grand Music Hall to accommodate 5,000. Complimentary tickets were sold on the street for $4.00 and $5.00. Four thousand programs were printed containing biographical sketches of the inventor.[11] His popularity seemed unbounded. He could not escape his Palm Sunday.

Said Tesla to the St. Louis audience, "The day when we shall know exactly what electricity is, will chronicle an event probably greater than any other recorded in the human race."[12] In Chicago, the dreams of Tesla were interwoven with the success of Westinghouse. Westinghouse was rewarded for his foresightedness in securing the World's Fair contract at any cost, for his company was now in line for consideration—and for ultimate success—in harnessing the power of Niagara Falls, through utilization of Tesla's polyphase system. Through Westinghouse was to come the realization of Tesla's boyhood dream.

Notes

1. *Electrical Engineer*, April 6, 1892, 350.
2. Ibid., August 31, 1892, 202.
3. *Engineering News*, March 5, 1892, 216.
4. Ibid., March 12, 1892, 240-241.
5. Ibid., May 26, 1892, 540.
6. Francis E. Leupp, *George Westinghouse*, (Boston, Mass.: Little, Brown & Co., 1918), 162-164.
7. Merrick Jackson, *George Westinghouse*, (New York, N.Y.: Mercer Publishing Co., 1958), 8.
8. Leupp, *George Westinghouse*, 167.
9. Interview with J. A. Knight.
10. John J. O'Neill, *Prodigal Genius*, (New York, N.Y.: Ives Washburn, Inc., 1944), 103.

11. Kenneth M. Swezey, "Nikola Tesla, Pathfinder of the Electrical Age," *Electrical Engineer*, September 1956, 5.

12. *Nikola Tesla, Lectures, Patents, and Articles*, (Beograd, Yugoslavia: Nikola Tesla Museum), i-iii.

George Westinghouse, early 20th century. *Joseph Gessford photograph, Library of Congress Prints and Photographs Division Washington, D.C., (LC-USZ62-93492).*

Arthur Brisbane, ca. 1906. *Library of Congress Prints and Photographs Division Washington, D.C., (LC-USZ62-112725).*

A Columnist's Viewpoint

The battle between scientists was not limited to science. The papers made such comparative headlines as "Our Foremost Electrician . . . Greater Even Than Edison." Arthur Brisbane, dean of Hearst's editorial writers, too, fell under Tesla's spell and was extravagant in his praise, even before the Niagara Falls project. Brisbane went to Delmonico's one evening to interview Tesla for a Sunday feature for *The New York World.* The two talked all night. Brisbane was surprised to notice that all but one of the lights had gone out and Mr. Delmonico's scrub women were coming in the feeble daylight to scrub the marble floor before his interview was finished.

In his article the reporter described the inventor as "the best electrician living . . . he is serious, he is earnest, and in all ways he commands respect." Brisbane continued:

> Every scientist knows his work and every foolish person included in the category of New York society knows his face. He dines at Delmonico's every day. He sits each night at a table near the window. When Ward McAllister, that strange contradiction of the theory that nature abhors a vacuum, wanders in, he sees Nikola Tesla with his head buried in the evening paper.
>
> Every foolish young man who cares for the law of gravitation only because it interferes with jumping fences, every foolish young woman who thinks that there is something new about her two-cent love affair, has seen this serious, owl-faced Servian eating his dinner and thinking about electrical vibrations. Nikola Tesla is almost the tallest, almost the thinnest, and certainly the most serious man who goes to Delmonico's regularly.
>
> He has eyes set far back in his head. They are rather light. I asked how he could have such light eyes and be a Slav.

> He told me that his eyes were once much darker but that using his mind a great deal had made them many shades lighter. I have often heard it said that using the brain makes the eyes lighter in color. Tesla's confirmation of the theory through his personal experience is important.
>
> . . . He has very big hands . . . his thumbs are remarkably big even for such big hands . . . This is a good sign. The thumb is the intellectual part of the hand . . . Nikola Tesla has a head that spreads out at the top like a fan. His head is shaped like a wedge. His chin is pointed as an ice-pick. His mouth is too small. His chin, though not weak, is not strong enough. His face cannot be studied and judged like the faces of other men, for he is not a worker in practical fields. He lives life up in the top of his head, where ideas are born and up there he has plenty of room. His hair is jet black and curly. He stoops—most men do, when they have no peacock blood in them. He lives inside of himself. He takes a profound interest in his own work. He has that supply of self-love and self-confidence which usually goes with success and differs from most of the men who are written and talked about in the fact that he has something to tell.

Brisbane was also surprised to find that mental calculations were relaxation for the genius. Their conversation ranged widely over a multiplicity of subjects. The reporter was amused at Tesla's description of the vibrations of a fly's wing which would be required to move it through the air. Tesla declared the fly to be the strongest living being in proportion to its size and said its wings moved about 25,000 times to the second. He calculated that this would make the fly use up about six hundred million times as much force as you might think to move his wings thus.

Then Tesla and the black-bearded restauranteer, Delmonico, entered into a friendly debate about whether or not there were man-eating sharks, and Tesla recalled the pike of the lower Danube which made swimming hazardous.

An engraving of this Sarony studio photograph of Nikola Tesla appeared in *The Century Illustrated Monthly Magazine*, February 1894. *Bain News Service photograph, Library of Congress Prints and Photographs Division Washington, D.C., (LC-DIG-ggbain-04851).*

Brisbane was eager to bring his interesting conversationalist to the point of his electrical theories. The two men discussed Tesla's hypothesis of light which, at the present, was his biggest undertaking. Tesla could not help revealing some disdain for Edison's incandescent lamp which he felt he could excel many times. The reporter concurred, saying, "compared with Tesla's his idea is as primitive as an ox-cart with two solid wooden wheels compared to modern railroading."

The scientist told Brisbane that the mystery of the light of the sun was the result of vibrations in 94 million miles of ether, separating us from the center of the solar system and predicted that he could produce here on earth vibrations similar to those which caused the sunlight and thus could give us light as good as that of the sun with no interference from clouds or other obstructions.

The reporter recalled some of Tesla's laboratory demonstrations in which he came out of his experiments a most radiant creature, with light flaming at every pore of his skin and from the tips of his fingers and from every hair on his head.

The two discussed this further until it became a staggering picture as Tesla described the five hundred trillion vibrations which occur in the ether every second to produce light. The scientist said, "All I have to do to duplicate the sunlight is to get this number of vibrations to the second with my machinery.... I have succeeded up to a certain point and am still at work on the task."

The interviewer was bewildered by the figure of five hundred trillion. Tesla tried to explain that:

> if a mass of metal as big as the Delmonico Restaurant, in which we sat, possessing ten thousand times the resisting force of the most finely tempered steel, should be caused to vibrate with one millionth of the rapidity of light-producing electric vibrations in ether, that mass of metal, ten thousand times harder than steel, would simply vanish into air like smoke. It would disappear into separate atoms too small to be seen and would never be heard of again.

Tesla told the reporter that electricity had no weight and therefore no opposition is offered to its moving backward and forward freely any number of times to the second. He said, "It is perfectly easy to prove that electricity weighs nothing. I will load you so full of electricity that you can't hold any more and then put you on the finest weighing machine, and you will not find one-thousandth part of an ounce added to your weight."

There was a bit of dry humor in this mysterious man as they talked of vibrations, and Brisbane suggested that, as a vibrator, electricity might meet with serious competition among modern statesmen. Telsa retorted, "No statesman could vibrate fast enough to be of any value scientifically."

The two discussed Tesla's rotating magnetic field which everyone was talking about but no one understood. The scientist explained that a magnet will seize a piece of iron and hold it firmly. Everybody knows that the magnet must use up force in holding that iron but of course as long as it holds the iron perfectly still that force is wasted. The piece of iron, if left alone, would stand still. There is no use in getting a magnet to make it stand still. But Tesla had found that he could get a magnet to use its force in such a way as to cause a piece of iron pipe to spin violently round and round. He could make a wheel at a distance from the source of electromagnetic force spin around with ten thousand horsepower, the basic principle in his plans for Niagara Falls.

Tesla hesitated to predict all the things which occupied his plans at the time, but he did confess that he expected to send messages through the earth without any wires and to transmit electrical force without waste. He explained the difference between voltage and amperage and that with this understanding and use of low amperage he could put a million volts of electricity safely through his body.

He went on to reveal an idea with which he had always been obsessed—that of relieving human slavery and lightening its work load. He predicted that he could solve the labor problem. Brisbane thought Mr. Debs might well consider this while languishing in his dungeon. Tesla predicted the push-button age when daily tasks could be done by simply pressing buttons.

The inventor manifested great pride in his Serbian origin and declared that the world would be surprised if it could understand the Slavic poets of his country. He declared that work killed no one, but that play could, and as for love and marriage, they interfered with success. In discussing food, he said with tongue in cheek that suppers are bad for one in New York but all right in Paris.

The two went on to a discussion of psychic phenomena and telepathy which the leading exponents of the day believed to be a sort of psychical electricity, Tesla considered the relationship pure coincidence, although he confessed to be amazed at the workings of the mind.

Brisbane questioned him as to the theory that there were a mere half-dozen fundamental laws governing the universe. Tesla countered, "I think that they could all be reduced to one."[1]

The two had an evening of mutual enjoyment, as was borne out by the five columns of copy under Brisbane's by-line in a Sunday section of *The World.*

As Tesla had revealed to the reporter, this was a great period of wireless experiment for him. He set up a transmitting station for sending messages through the earth and dotted his receiving stations at points in the city. He planned to reach out to greater distance by setting up a boat on the Hudson.

The laboratory at 33-35 South Fifth Avenue (now West Broadway) literally vibrated with activity. The neighbors were increasingly curious and suspicious of what was going on. Strange lights and weird noises behind locked doors did nothing to allay their fears. From time to time visitors pounded on the entryway and were quietly admitted and the doors again locked behind them. The residents of the area had no concept of what this man might be doing.

If they had been able to peer behind the drawn curtains they would still have been mystified by the sight of a maze of wires and the weird mechanical contrivances. The genius and his assistants were working on shadowgraph pictures, forerunner of X-ray principles. Tesla laid no claim to X-ray, but he achieved pictures with similar results. He took photographic images through a skull, forty feet from his light tubes. In December of 1895, when Professor Wilhelm Roentgen in Germany announced his discovery of X-rays, Tesla forwarded to the German scientist pictures which had been produced with his shadow-graph technique. Roentgen was naturally interested in the pictures and in how Tesla had obtained them. Tesla laid no claim to discovery but had proceeded merely out of scientific curiosity. He was invited to write an article about the Roentgen discoveries in *The Electrical Review,* an article in which he gave full credit to the German scientist. Tesla discussed the nature of the rays and brought out his own observations and conclusions as a result of his experiments. He related the application of the discoveries to surgery and gave his own theories of danger to the skin in use of the rays.[2]

Wilhelm Roentgen discovered X-rays and exposed the first X-ray photograph in December 1895. Nikola Tesla was experimenting with shadowgraph pictures at about the same time. *Bain News Service photograph, Library of Congress Prints and Photographs Division Washington, D.C., (LC-DIG-ggbain-05679).*

Century Magazine sent a symposium of comments on the Roentgen discoveries to Tesla, asking him to include his own ideas to be incorporated with those of Edison, Elihu and Sylvanus Thomson, and others. His reply to Robert Underwood Johnson, the editor, was that Johnson would do well to throw the symposium in the waste-basket because knowledge of the rays was developing so rapidly that by the time the article could be published it would be a burlesque. He felt sure that all the contributors would be grateful to have their erroneous views suppressed—that premature expressions without substantiation would be worthless.[3]

Simultaneous with his work on the shadowgraph Tesla was continuing his development of teleautomatics and perfecting an electron microscope. His carbon button lamp, for which there seemed no reason or place, held principles of the atom-smasher-to-be. A myriad of his heterogeneous ideas had embryonic form in the South Fifth Avenue laboratory. The 13th of March, 1895, proved disastrous to Tesla. He had retired to his room for a few hours of rest when he was aroused by someone beating on his door to tell him that the laboratory was on fire. The flames had started in the lower part of the building and the entire structure was gutted. The upper floors collapsed. His equipment was beyond salvage. Records of his World's Fair achievements had gone up in smoke along with the nearly completed model for his first wireless system demonstration. Not one article or memento was saved. The catastrophe was a complete shock to the inventor, who was always so careful about fire. His impracticality in business exacted its penalty. There was no insurance. The building had been the property of the Tesla Electric Company. A. K. Brown and Charles Peck were the losers, along with Tesla.

He could depend on money from his German patents to provide his living expenses, but it would be wholly inadequate for the building and equipping of a new laboratory. Copper bars the size of 2x4s ran into money. Bystanders expressed their regrets, particularly for the irreplaceable data. That loss was of little concern to Tesla. His photographic mind retained the record of his experiments in minute detail.

Charles Dana, writing in *The New York Sun,* said:

> The destruction of Nikola Tesla's workshop with its wonderful contents is something more than a private calamity. It is a misfortune to the whole world. It is not in any degree an exaggeration to say that the men living at this time who are more important to the human race than this young gentleman can be counted on the fingers of one hand; perhaps on the thumb of one hand.[4]

The Scientific American of March 23, 1895, described Tesla's loss and declared that it could not be reckoned in dollars and cents. They expressed dismay at the destruction of several nearly completed inventions, but predicted that it might also adversely affect Tesla's health, already bordering on exhaustion from overwork. The article also mentioned the lectures which Tesla had given before European and American scientific bodies, then went on to list honorary degrees from Yale, Columbia, and other noted universities. He was listed as the controlling engineer in the Niagara Falls Power Company.[5]

Friends offered condolences, but ironically it was the enemy forces who came forth with financial backing so that Tesla might reopen a laboratory. At 46 East Houston Street, with a grant of $40,000 from Edward Dean Adams, head of the Morgan group, Tesla was able to reopen a laboratory within four months. Furthermore, the Morgan group offered other financial assistance along with alliance with General Electric, a group which could have spelled financial security and prestige.[6] There may have been memories of unhealed wounds, unfilled promises, and the stinging practical jokes in the Edison Company and the remnants of incompatibility with workers even in the friendly Westinghouse Corporation. Tesla must have been fully aware of his own "allergy" for experts. He turned down the sure thing for the uncertainties of a future based on his own ability.

Notes

1. Arthur Brisbane, "Our Foremost Electrician," *The World*, July 22, 1894, 17.

2. *Electrical Review of New York*, December 2, 1896, 276.

3. Microfilmed letters, Robert Underwood Johnson to Nikola Tesla, March 10, 1896, and Nikola Tesla to Robert Underwood Johnson, March 12, 1896, Library of Congress.

4. Charles Dana in *New York Sun*, March 13, 1895.

5. *Scientific American*, March 23, 1895, 185. Bibliography of Dr. Nikola Tesla, published by The Tesla Society, Minneapolis, Minn.

6. John J. O'Neill, *Prodigal Genius*, (New York, N.Y.: Ives Washburn, Inc., 1944), 124-125.

The Niagara Falls Hydraulic Power & Manufacturing Company, founded in 1853, began commercial generation of direct current electricity, on a small scale, in 1882. This photograph, taken between 1880 and 1897, of the Canadian Horseshoe Falls shows the mist that was present prior to exploiting the incredible hydroelectric potential for alternating current power generation. *Detroit Publishing Company photograph by William Henry Jackson, Library of Congress Prints and Photographs Division Washington, D.C., (LC-D4-4026 A).*

Niagara Falls & the Termination of the Battle of the Currents

Niagara Falls, the one-hundred-sixty-four foot torrent that wasted constantly over the gorge, had long been studied as a power source. A seventeen-man commission had made a careful study of practical means of utilizing this natural force. Lord Kelvin, the capable British head of the commission, had established himself by his own achievements in the field of electrical engineering. He had been instrumental in the development of telegraphy and the laying of the Atlantic cable. Among his inventions was a mirror galvanometer used in cable signaling and a siphon recorder. He had probed the mysteries of magnetism and held the chair of Natural Philosophy at the University of Glasgow. Cambridge and Dublin mathematical journals named him as editor. He had been given the title of Baron Kelvin of Larg in 1892.[1] As a recipient of many European honors, he held the confidence of leading engineers in the world. Also he had been a firm believer in the advantage of Edison's direct current. But in 1893 he began to change his mind. The World's Fair in Chicago, which had been so successful as a proving ground for the Westinghouse current, plus other successes which were being chalked up to alternating current experiments, were becoming factors for consideration.

The city of Buffalo was aware that the utilization of power through alternating current had staggering possibilities. Engineers began to toy with the extravagant idea of transmission of power from the falls—for the difficult distance of twenty-two miles, a feat completely impossible by direct current. If this miracle could be assured, even New York City might be lighted in this manner. There was the precedent in Germany, now, for at the Frankfort Exhibition, a 100 horsepower motor, by means of a three-phase alternating current set-up, had utilized power transmitted from a generator in Lauffen, 108 miles away.[2] This had signaled unlimited possibilities.

More than seventeen plans had been submitted for harnessing Niagara Falls, by 1891, but the commission had been disbanded without accepting any of them. Twenty thousand dollars had been offered as an inducement for feasible plans but had not been collected. Now, the use of alternating current was almost a certainty.

The International Niagara Falls Commission: Prof. Eleuthere-Elie-Nicolas Mascart, Prof. William Cawthorne Unwin, Lord Kelvin, Dr. Coleman Sellers, Col. Theodore Turrettini. *From* Cassier's Magazine, *courtesy of Fran Kiester.*

Proposals and bids were asked on a power system of three generating units each of 5,000 horsepower. On October 24, 1893, Westinghouse was awarded the initial contract, and his victory in the "War of the Currents" was conceded. General Electric found it necessary to secure a license for the use of Tesla patents. Westinghouse completed his powerhouse in 1895 and the residents of Niagara Falls area turned on the lights on April 20th.[3] The installation of the three generators with a combined capacity of 15,000 horsepower became the supreme electrical engineering feat of the time.

In 1896 the General Electric Company completed the transmission and distribution system to Buffalo. Professor Charles F. Scott, Professor Emeritus of Electrical Engineering at Yale University, was jubilant in his praise for Tesla's polyphase system which was being used. He declared it to be the most tremendous achievement in engineering history.[4] Buffalo became a city electrified. A maze of trolleys transported the workers to the expanding mills and mushrooming communities. The great electro-metallurgical and electrochemical industries boomed. Cheap power changed the face of industry. One of the first companies for Niagara power was the Pittsburgh Reduction Company, later to become the Aluminum Company of America. Another customer was Dr. Edward Goodrich Acheson, who was working with calcium carbide. He was able to found a giant industry which included carborundum and calcium cyanide for nitric acid explosives and fertilizer. In less than a generation the largest electrochemical community in the world had been built up around Niagara power.[5]

Westinghouse installed seven additional generating units, bringing the output to 50,000 horsepower. Later, a second equivalent powerhouse using eleven generators was installed by General Electric. The current was on its way to New York City. By 1903 practically all other stations would adopt the Tesla system.[6]

Tesla received congratulations and honors from friends and admirers everywhere upon completion of the Niagara Falls project. His closest friends, the Robert Underwood Johnsons, rejoiced in his success and on November 17, 1896, Johnson sent Tesla a facetious note. "Dear Mr. Tesla," he wrote, "It may seem presumptuous in a stranger to address you, but Mrs. Johnson, my wife whom you may not remember . . . cannot refrain from uniting with me in congratulations."[7] The missive was signed with the nickname "Filipov," which Johnson so often used in their correspondence.

The Niagara Falls and Conduit Company knew well where the honor for the patents involved in their achievement lay. They decided to honor Tesla at a celebratory dinner at the Ellicott Club on January 12, 1897.

The speakers of the evening paid glowing tribute to the inventor, stating that perhaps none was more deserving to be honored and that his electrical researches and practical accomplishments had been the talk of the world. Tesla was the star for the evening, but his speech was characterized by great humility. He paid tribute to others in his field, including Lenard, Roentgen, and Lord Rayleigh. He proclaimed his pride in being a citizen of the United States and praised this country's share in progress. He spoke of the influence of the artist who might become a physician, an electrician, an engineer, a mechanist, or even a financier, yet, even in seemingly unrelated fields, continue to be the artist. He gave gracious praise to Edison and said, "Had he not done anything else beyond his early work in incandescent lighting, he would have proved himself one of the greatest benefactors of the age."

He spoke of electricity as a poet might speak, repeating his thesis that the scientist is also an artist. His whole life had been a quest for the mysteries of electricity and on this occasion he expressed his belief that this force was tied up with the mystery of life and perhaps even the explanation of creative force.

> We know that light, heat, electrical and magnetic actions are closely related, not to say identical. The chemist professes that the combination and separation of bodies he observes are due to electrical forces and the physician and psychologist will tell you that life's progress is electrical. Thus electrical science has gained a universal meaning, and with right, this age can claim the name, "The Age of Electricity."[8]

In a former lecture, he had concluded with the words, "I think I know what electricity is." But on this night, he stated boldly, "I wish much to tell you on this occasion—I may say I actually burn for desire of telling you what electricity *is,* but I have strong reasons which my co-workers will appreciate. . . . I shall not dwell on this purely scientific aspect of electricity."

His emphasis, that evening, was on the hopes brought about by the Niagara achievement, which he called "a monument worthy of our scientific age, a true monument of enlightenment and of peace." He predicted the end of barbarous methods and the relief of want and suffering.

"Power is our mainstay," he declared, "the primary source of our many-sided energies, and the greatest criminals of all are the voluntarily idle." He concluded,

> With ideas, it is as with dizzy heights. At first they cause you discomfort and you are anxious to get down, distrustful of your own powers; but soon remoteness of the turmoil of life and the inspiring influence of the altitude calm your blood and your step gets firm and sure and you begin to look—for dizzier heights.[9]

So he explained the thought process of inventive genius.

Once upon a time, a dream-hungry Serbian lad opened the pages of an old book and turned to the steel engraving of Niagara Falls and said, "Someday I will harness those falls!"

Tonight, in his moment of triumph, he was able to say, *"Fait accompli."*

Notes

1. *World Book*, vol. X, (Chicago, Ill.: Quarrie Corp., 1947), 4128.

2. Kenneth M. Swezey, "Nikola Tesla," *Science*, May 16, 1958, vol. 127, no. 3307, 1151.

3. Merrick Johnson, *George Westinghouse*, (New York, N.Y.: Mercer Publishing Company, pamphlet), 9.

4. John J. O'Neil, *Prodigal Genius*, (New York, N.Y.: Ives Washburn, Inc., 1944), 106-107.

5. Swezey, "Nikola Tesla," 1152.

6. Kenneth M. Swezey, "Nikola Tesla, Pathfinder of the Electrical Age," *Electrical Engineering*, September 1956, 2.

7. Microfilm letters, Johnson to Tesla, November 17, 1896, Library of Congress.

8. *Nikola Tesla, 1856 – 1943, Lectures, Patents, and Articles*, (Beograd, Yugoslavia: Tesla Museum, 1956), A-102-A-104.

9. Ibid., 108.

Nikola Tesla, "Showing the Inventor in the Effulgent Glory of Myriad Tongues of Electric Flames After He Has Saturated Himself with Electricity." *From* The World, *New York, July 22, 1894.*

Experiments in Resonance & Teleautomatics

Tesla became ever more wary of divulging all of his plans until patents were issued. He was learning that there were no ethics where invention and money were involved. Yet he had a completely revolutionary experiment impending which was sure to rock the world. Reporters, to whom he gave broad hints, were in an equal quandary how much to report. They dared not ignore a prophecy from a man who had a way of making his dreams come true. Neither could they be premature in their enthusiasm. It was sometimes a choice between being scooped and being ridiculed.

A few publications took their chances on being ridiculed and labeled Tesla's hints sensationalism. *The Electrical Engineer* insisted that Tesla had gone far beyond the possible ideas he had put forth and that he had behind him a trail of beautiful but unfinished inventions. They proceeded to quote from a letter in *The New York Sun* which said that Tesla had claimed his ability to send projectiles at command, arrest them in flight, and call them back to explode at will. The editors of *The Electrical Engineer* maintained:

> When we are expected, wide awake and in our sober senses, to accept in silence such an utterance . . . or that which describes as "a possibility" the operation of a distant torpedo boat by the mere exercise of the will, we refuse point blank and are willing to face the consequences. Our past admiration of Mr. Tesla's real, tangible work is on record, and stands; but we draw the line at such things as these.[1]

Scientific American also deplored the sensational interviews and scourged Tesla for allying himself with the mongers of yellow journalism.

> Judging from the comments of the scientific and technical press, we are not alone in our expressions of regret that any one of Mr. Tesla's undoubted ability should indulge in such obvious and questionable self-advertisement. That the author of the multiphase system of transmission, should, at this late day, be flooding the press with rhetorical bombast that recalls the wildest days of the Keely motor mania is inconsistent and inexplicable to the last degree.[2]

On September 2, 1897, he secured his patent protection, and it was the following year that he refuted the critics.

Scientists were crowding each other to the wall with wireless inventions, and already Edison, Bell, Faraday, and Hertz had made progress in the field in some form. The scientific world was bursting with pyrotechnical displays, but no one presented a more graphic demonstration than did Tesla in his week-long display at Madison Square Garden. Important personages came by invitation. The capacity of the building was strained to the limit and crowds were turned away. Any demonstration of Tesla's was bound to be thrilling. Each night, at the time appointed, the doors were closed and Tesla appeared in full dress. He announced that they were about to see a boat piloted by remote control, using a borrowed brain—first guided by his will and then by those of some of the members of the audience. A huge tank had been constructed in the center of the auditorium, filled with water on which floated a metal covered boat several feet long and equipped with electric lights. By means of radio principles Tesla's every wish was performed by the boat without connecting wires and later on the audience took its turn in giving commands.

To Tesla, the basic idea was quite simple. It was based on his study of his own physical reactions to certain stimuli. He had come to think of the human mechanism as another machine. Why could one not build an automaton which would respond to external commands? And if it could follow human wishes, why not consider its further application in warfare? Its destructiveness would be unlimited and thus would tend to end wars.

The audience was quick to grasp the significance of such power with their involvement in the Spanish American War.

The demonstration at Madison Square Garden, dramatic as it was, revealed only a part of what the patents contained. Included were specifications for a torpedo boat without a crew, involving a motor with storage battery to drive the propeller, smaller motors and batteries to operate the steering gear, and still others to feed electric signal lights with compressed air motors to raise or lower the boat in water. It was to be equipped with six fourteen-foot Whitehead torpedoes, placed vertically in two rows. As one was discharged another would fall into place to be fired through a tube. He estimated that such a boat could be built for $48,000 to $50,000 or less and that all that was necessary was to wire it properly. All power could be stored in the boat itself with vibrating sound to carry messages without any connecting wires. Tesla's plan involved the use of several submarine destroyers aboard the ship which could be launched from it. The director could know the exact position of the torpedo boat by means of two projected masts to be lighted with hooded lights at night. "These little destroyers," the inventor said, "could attack and destroy a whole armada—destroy it utterly in an hour, and the enemy never have a sight of their antagonists or know what power destroyed them."[3]

With the graphic demonstration at Madison Square Garden the public was entertained and intrigued and could grasp a part of the potential uses of a mind-powered machine, but they were still skeptical. They were inclined to believe that his machine would prove impractical. The idea of the kind of warfare which might be let loose with such thinking was unbelievable, unthinkable, and they preferred to hide their heads in the sand. With each of Tesla's announcements during that week they tittered nervously, afraid to believe and afraid to doubt. Although Tesla continued to stress the preventive aspects of his inventions in warfare, the world was, as usual, war-minded, and unable to envision a time without conflict.

There were some, however, who grasped the scope of the Tesla mind. One reader of *The Electrical Review* wrote the editor concerning the inventor; "He has simply gone after tarpon when others have been content with sprats."[4]

But Tesla's "impossible" plans did not end here. As a further extension of his torpedo boat controlled by the human mind, he was working toward an automaton which would have a mind of its own, that is, one which would be able to operate independently of human will and to perform as if it had intelligence. He startled the public with such statements and added that his automaton

> will be able to follow a course laid out or to obey orders given far in advance; it will be capable of distinguishing between what it ought and what it ought not to do, and of making experiences, or, otherwise stated, of recording impressions which will definitely affect its subsequent actions. In fact, I have already conceived such a plan.[5]

This may well have been an embryo cybernetic machine.

Nor was this the limit of his plans. He was convinced that the principle contained still further application and could be employed by any machine that moves on land or in water or in the air and further said:

> . . . it affords a means of absolute controlling . . . all the innumerable translatory movements, as well as the operation of all the internal organs, no matter how many, of an individualized automaton.[6]

Kenneth M. Swezey, writing for *Science,* indicated that one of Tesla's original model boats did utilize far more intricate control apparatus than was evident even in the patent No. 613,809. The writer said:

> To protect inventions he was not yet ready to disclose, he outlined only the basic idea in his specifications; besides, at the time his application was filed, utilization of the invention was seriously being considered by the United States Navy for help in our war with Spain. One of the features not revealed was a system to prevent the interference by means of coordinated tuning devices responsive only to a combination of several radio waves

> of completely different frequencies, Another was a loop antenna which could be completely enclosed by the copper hull of the vessel; the antenna would thus be invisible and the vessel could operate completely submerged.[7]

It is evident from this, that dramatic though the Madison Square Garden demonstrations were, they revealed only a small fraction of the tremendous ideas buzzing in the Tesla brain.

The small waves vibrated by his moving boat stirred responses in his own mind. Here, in these vibrations, was power that might be utilized. The perfect tuning of his radio instruments caused him to reflect on the laws of resonance. Guglielmo Marconi's short waves and his own long waves were matter for speculation. After all, music was only one instance of science based on vibration. The world was full of examples which he could put to use. He explored the strange relationship by which articles move when tuned to the resonance of another vibrating object. The therapeutic effects of massage demonstrated the beneficial uses of vibration. Tesla pondered on the old truth that men marching in step upon a bridge could affect its collapse by maintaining the rhythm of their steps. Resonance was an untapped source of power.

According to A. L. Benson, a reporter, Tesla ordered for his Houston Street laboratory a steel link two feet long and two inches thick, of the best quality steel, strong enough to bear the weight of a hundred tons. To this he fastened a small electric vibrator. As the writer described it, the vibrator

> was no larger than an alarm-clock, so constructed that frequency of vibrations could be altered at will. He set the vibrator in "tune" with the link. For a long time nothing happened—vibrations of link and machine did not chance to coincide, but at last they did and the great steel link began to tremble, increased its trembling until it dilated and contracted like a beating heart—and finally broke.

> Sledge hammers could not have done it; crowbars could not have done it, but a fusilade of taps, no one of which would have harmed a baby, did it. Tesla was pleased. He had learned something. He wanted to learn more. He put his little vibrator in his coat-pocket and went out to hunt a half-erected steel building. Down in the Wall Street district, he found one—ten stories of steel framework without a brick or a stone laid around it. He clamped his vibrator to one of the beams, and fussed with the adjustment until he got it.
>
> Tesla said finally the structure began to creak and weave and the steel-workers came to the ground panic-stricken, believing that there had been an earthquake. Police were called out. Tesla put the vibrator in his pocket and went away. Ten minutes more and he could have laid the building in the street. And, with the same vibrator he could have dropped the Brooklyn Bridge into the East River in less than an hour.[8]

This was a shocking enough revelation of the powers of vibration, but the inventor startled Benson still further by insisting that he could split the earth in the same way, "split it as a boy would split an apple—and forever end the career of man." Tesla explained it clearly by saying:

> The vibrations of the earth have a periodicity of approximately one hour and forty-nine minutes. That is to say, if I strike the earth this instant, a wave of contraction goes through it that will come back in one hour and forty-nine minutes in the form of expansion. . . . The earth, like everything else, is in a constant state of vibration . . . continually contracting and expanding.
>
> Now, suppose that at the precise moment when it begins to contract, I explode a ton of dynamite. That accelerates the contraction and, in one hour and forty-nine minutes, there comes an equally accelerated wave of expansion. . . . Suppose this performance to be repeated, time after

> time. Is there any doubt as to what would happen? There is no doubt in my mind. The earth would be split in two. For the first time in man's history he has the knowledge with which he may interfere with cosmic processes.[9]

When the reporter inquired how long Tesla thought it would take him to split the earth in two, he said he didn't know, perhaps months, or even a year or two, but Tesla went on:

> in a few weeks I could set the earth's crust into such a state of vibration that it would rise and fall hundreds of feet, throwing rivers out of beds, wrecking buildings, and practically destroying civilization. The principle cannot fail. . . . There may be flaws in this theory. Others may see them. I don't.[10]

Benson considered the possibility stimulating to the imagination, but was convinced that the improbability of its application was sufficient to allay one's fears.

Friends of Tesla claim that there was another dramatic experiment with resonance in his Houston Street laboratory. On this particular occasion he fastened the small, innocent-looking device to a rigid pole. The surrounding neighborhood was a conglomeration of tenements populated by Italians and crowded by the workers in garment-factory sweatshops and nearby Chinatown. The quarter was puzzled and fearful of goings-on in Tesla's loft room.

Confident that the building was set on solid granite, Tesla allowed the vibrations to build higher and higher. Apparently they fanned out from the laboratory in an unpredicted manner, through the sand which underlay the building's foundations. The sand conducted the waves with earthquake intensity. The story is told that the police headquarters in Mulberry Street felt the impact even before the scientist was aware of it in his own laboratory and that prisoners became terrified of an earthquake. Furniture in the police station danced about the room and plaster fell from the ceiling. The location of chairs proved disconcertingly uncertain to those seeking a place to sit down. The officers, always apprehensive of the goings-on at Tesla's laboratory, were dispatched to the scene to investigate. When they reached the site, they found Tesla was smashing the machine with a sledge hammer. Turning off the current would not have been sufficient, as the resonance would have continued to mount, perhaps to disastrous proportions.

It is believed that Tesla never gave out the secret of this small motor, but that in the thirties before the outbreak of the war, when he was lecturing, he declared that he could apply these same principles for the detection of submarines. His system of telegeodynamics, he asserted, could be used to locate ore deposits.[11]

In reviewing the subject, Tesla said that the effects of the telegeodynamic oscillator were so powerful that he could take down the Empire State Building in a very short time. He described the small mechanical vibrating device which he would employ as being so small it could be concealed in a pocket. By attaching this device to a part of the building and allowing the resonance to build up for twelve to thirteen minutes, he predicted that the rivets would be loosened, and the outer coating would crumble and the steel structure collapse.[12]

So go the stories of the years in which Tesla experimented with teleautomatics, telegeodynamics, and all his theories of resonance. There are those who will say there is no police record of this vibration episode on Houston Street and others who feel that the police probably considered Tesla crackpot but harmless and not worth alarming the neighborhood. There are some who think it may be pure myth. But there are others who insist that the laws of resonance are based on principles that the Israelites unknowingly put into operation, not in thirteen minutes, but in seven days and seven nights, with a ram's horn and a shout!

Notes

1. *Public Opinion*, vol. xxv, no. 22, December 1, 1898, 684-685, reprinted from *Electrical Engineer*.

2. *Scientific American*, November 26, 1898.

3. "Torpedo Without a Crew," *Current Literature*, February 1899, 136-137.

4. *The Electrical Review*, November 9, 1898, vol. 33, no. 19, 279.

5. Nikola Tesla, "The Problem of Increasing Human Energy," *The Century Magazine*, June 1900, 175-211.

6. Ibid.

7. Kenneth M. Swezey, "Nikola Tesla," *Science*, May 16, 1958, vol. 127, 1156.

8. A. L. Benson, "Nikola Tesla, Dreamer," *The World Today*, February 1912, 1763-1767.

9. Ibid.

10. Ibid.

11. John J. O'Neill, *Prodigal Genius*, (New York, N.Y.: Ives, Washburn, Inc., 1944), 164-165.

12. Ibid.

Nikola Tesla and Marion Crawford in Tesla's laboratory with the fluorescent light. *From* Remembered Yesterdays, *by Robert Underwood Johnson, Little, Brown & Co., courtesy of Fran Kiester.*

Nikola Tesla's Colorado Springs experimental station. The front of the laboratory faced west, with a full view of Pikes Peak. According to Tesla, Dickson D. Alley made this photograph on an 11x14-inch glass plate one "afternoon before sunset," late December 1899. The Union Printers Home is seen in the distance. The handwritten text reads, "Experimental station in another phase of development. Discovery of 'stationary waves' made July 3, 1899." *Courtesy Nikola Tesla Museum, Belgrade, Serbia, (006-04-009).*

Tesla's Laboratory in Colorado Springs

Tesla was outgrowing his New York laboratory in 1899. He longed for space and privacy to carry out certain experiments which would be dangerous in urban areas. It was his old friend, Leonard Curtis, who invited him to come to Colorado Springs. Curtis had been patent attorney for both Tesla and Westinghouse during the tense "War of the Currents." He had broken his health in those strenuous times and had come to Colorado to recuperate and rest. In Colorado he became associated with power developments, invested is coal mines near Colorado Springs, became affiliated with the Colorado Springs Electric Company, and added his name to the legal firm of Hall, Preston and Bobbitt. He accepted a position of director on the board of the Colorado Title and Trust Company and bought one of the big houses in the 1400 block of North Cascade where he entertained Little London's millionaires. He was recuperating but not exactly resting.

"Come to Colorado Springs," wrote Curtis to Tesla, "and I can get you all the land you want, rent free from the Colorado Springs Company." He was referring to the prairie pasture east of the Deaf and Blind School. "And all the electricity you need free from the Colorado Springs Electric Company," he added. In return, he felt that Tesla would be a charming visitor who would contribute to the prestige of the region.

John Jacob Astor had $30,000 he was anxious to invest and was willing to lend it to Tesla at this time. William Crawford, of the drygoods firm of Simpson and Crawford, lent an additional $10,000. Tesla was delighted with the location offered to him. He was anxious to observe and experiment with lightning at high altitude where conditions would be favorable for his new experiments.

It was May 17th that Tesla arrived in Colorado Springs to be met by Curtis and to be whisked off to lunch at the El Paso Club, Little London's equivalent to the haunts of Ward McAllister's 400. He was tired and anxious to check into a room at the Alta Vista Hotel. He chose room 207. It had two advantages. It was only one flight up, and one look at the primitive elevator convinced him he would rather climb the stairs. A private bath and pleasant view of Pikes Peak from his west window were added inducements. Not to be overlooked was the number 207. It was divisible by three—an old superstition about the number three that could not be overlooked.

John Jacob Astor IV provided Nikola Tesla with financial backing for his laboratory and research in Colorado Springs. Astor was a man of great wealth, having built the Astoria addition to the Waldorf-Astoria Hotel in 1897. Astor drowned on the maiden voyage of the *R.M.S. Titanic* on April 15, 1912. *Bain News Service photograph, Library of Congress Prints and Photographs Division Washington, D.C., (LC-DIG-ggbain-03821).*

In the corridor, eager reporters met him with a barrage of questions. They wanted to know why he came, what he was going to do, and when he intended to do it. Tesla answered that he intended to send a wireless message from Pikes Peak to Paris. He was also planning an exhaustive research in regard to unsolved phases of wireless experiments and said that he wished to investigate electrical disturbances at this altitude.

When the reporters asked him if he intended to send messages from peak to peak, he evidenced some irritation, saying that he was here for serious work and not for stunts or pleasure. With weariness, he added, "I am very busy and life is short and there is a great deal to be done." It was one of those moments when he was becoming increasingly aware that although his ideas were unlimited, their execution was circumscribed by time. One life span, no matter how long, would not be sufficient.

Nikola Tesla arrived in Colorado Springs on Wednesday, May 17, 1899. He traveled from Denver on the Union Pacific, Denver and Gulf Railway, arriving at 1:00 p.m. The "New" Alta Vista Hotel (pictured), 118 North Cascade Avenue, was Tesla's home while he resided in Colorado Springs during 1899–1900. He stayed in room 207 in the newly expanded hotel. An 1898 advertisement for the Alta Vista stated, "Rooms superior to any hotel in the city . . . All modern improvements. A table that has not superior. . . . Rates $2.50 to $4.00. Special by the week." The Alta Vista was razed during 1963. *From Special Collections, Pikes Peak Library District.*

Tesla begged to be excused—said he must rest. He went into a contemplative state, unconsciously excluding the trivia of the world and the reporters.[1]

He turned and climbed the stairs to his room after taking another quizzical look at the elevator. He was not hesitant to play with a million volts of electricity but he drew the line at some elevators.

The hotel was famous for its cuisine, and Tesla was pleased by the abundance of clean linens. In the parlor he had seen Hoyt Stevens' pretty young daughter Julie playing the piano. Julie was a gifted girl and smart as a whip. She carried a little autograph album in her pocket at all times. She was not long in getting the autograph of the mysterious stranger in 207. Tesla enjoyed hearing Julie's music lessons. He made a mental note to buy a book of Schumann for her and to urge her to learn "The

Prophet Bird." Fritz Lowenstein, Tesla's assistant, would like to hear her, too, when he arrived. Fritz enjoyed music. Together, they would show Julie exactly how to play it.[2]

The maid service at the hotel was good. No one objected when Tesla demanded extra changes of bed linen and eighteen clean towels a day in his room. Mrs. Stevens told the maid, "If he wants them, let him have them. It is good to have a clean gentleman in the room." She did not understand that this genius had dark corridors of compulsions and phobias. This germ phobia, which caused him to wash his hands continually, had obsessed him since childhood and the condition grew increasingly demanding with the passage of years.

Tesla was soon absorbed in contemplation of the huge undertaking which lay ahead of him. His responsibilities would be doubled because he would have to manage to direct each detail, both in the new location and in the laboratory in New York. Every bit of equipment to be used in Colorado Springs would have to be made in the east by carefully trained technicians. Because he was pioneering in a new field, there would have to be perfect coordination of the two laboratories. Lucky that he had such a capable man as George Scherff in charge of New York operations to relay instructions, look after his patent rights, report on daily work, and look after private business as well.

A report from Scherff a few days after his arrival assured him that they were proceeding full speed ahead at home, preparing drums, varnishing balloons, grinding the necks from several hundred bottles, and working with bearings. There would be nickel and aluminum chips ready soon, and five grades of silver filings to send. They were working with breaks and coherers. Clock works, too, would be ready to ship. Tesla was satisfied with the progress. He urged Scherff to get Lowenstein on the train to Colorado as soon as possible.[3]

Leonard Curtis was anxious to introduce his famous friend to the elite of Little London and some of Colorado's notables, so he planned a dinner party for May 23, at the El Paso Club. Twenty men were the selected guests invited to have the privilege of hearing this genius who had spoken before scientific royalty of Europe and America. One of the honor guests of the evening was Governor Charles S. Thomas, who spoke briefly to acknowledge the pleasure of the occasion. The guests waited eagerly to hear Tesla, who told them of his early life, his latest successes, and his ultimate plans for wireless telegraphy. His listeners were convinced that he was a peer in his profession, but they were confused about what he intended to do in Colorado. They were surprised by his retiring manner. They were not fully aware that greatness was among them.

Mrs. Gilbert McClurg (under the nom de plume of Desire Stanton), wife of the secretary of the chamber of commerce, reported for the newspaper *Mountain Sunshine*. She noted that Tesla was in his happiest vein at the dinner. She recalled:

> . . . in the first years of 1800, Lieutenant Pike had camped at the foot of this great "Mexican Mountain" — that far, blue cloud.[4]

Now that the last years of the same century were drawing to a close, the reporter pointed out that another discoverer was camping at the foot of Pikes Peak, "armed with a more mighty key than Poor Richard."

Tesla had been asked about his proposed laboratory, and had revealed his plans for a large building, about 50x60 and 18 feet high. It was to be crowned with a huge platform, made according to his own specifications. The location had been selected so that it would be in a central position for telegraphy from mountain to plain and from plain to mountain.

Tesla had warmed to the group and grown so daring as to predict that he would communicate with Mars. This statement had not been wholly accepted by his audience, yet this had been the kind of night when ideas get out of bounds. Tesla had told them of his plans to erect two terminal stations in London and New York.

Somehow the conversation had brought in a discussion of Edward Bulwer-Lytton's novel *The Coming Race,* with its use of "Vril power." There had been an explanation of automata, governed by transmission of circumstances upon the eye.

Tesla had ventured to predict that Colorado might well be the location of the first exhibition of "Vril power." For those who had not read the book, he explained that the writer had conceived a "Vril" staff by which power could be arrested or transmitted. An account of the boat which Tesla had controlled with wireless means at Madison Square Garden had also been a feature of evening entertainment.

Pikes Peak Avenue, with streetcar tracks running down the middle and flanked by electrical power lines, runs west from Knob Hill toward Colorado Springs, mid 1890s. Nikola Tesla's Colorado Springs laboratory was located near this vantage point and this view would have been a familiar one to Tesla. *Special Collections, Pikes Peak Library District (001-0624).*

Desire Stanton closed her report of the dinner with the following words:

> All who were present at the banquet, felt, if it may be stated without bathos, that they had soared toward the empyrean for a glimpse of glory that excelleth, the man in the midst of fire, the controlling power,
> "Half spirit and half bird,
> And all wonder and a wild desire."[5]

It was agreed that the twenty guests left the El Paso Club that evening with a sense of exhilaration, having been lifted momentarily from the boundaries of small-world thinking. Although they had been unable to comprehend the magnitude of Tesla's projects, they had been ignited by the fire of his excitement. Tomorrow they might make a more rational evaluation of the man, and the sheen of magic might disappear. They would probably be conservative Little London again, but for a moment their eyes had beheld horizons—unlimited.

It was a happy circumstance when Tesla secured Joseph Dozier to agree to build the big laboratory for him. There were to be receiving stations also dotted over various parts of town. Old Mr. Dozier was just enough of a visionary to understand Tesla whose great inventions had been built on a vision. Both men had fantastic dreams which had led to the rejection of the dreamers. Both were aware of the vast deposits of ore in the surrounding mountains. Both attributed the violent electrical storms of the region to the magnetic attraction of the ore deposits. Dozier had certain theories about gold in Cripple Creek which he thought might lie as deep as the Colorado Springs level. No one believed the old man, but Tesla listened and swapped equally unaccepted theories. Tesla told his friend of ball lightning which he had seen at the Bluffs for the first time. Dozier clued him in on lightning behavior at high altitudes.

Occasionally both men discussed matters that bordered on the occult. Tesla did not dare to speak of such things openly with many people. Always he denied taking any stock in mysticism. It was a strange world that scoffed at so many mysteries which he could explain. With Dozier he could let himself go. Even though he kept asserting that all mystery and miracles were the result of scientific truth—he was impatient to discover the key. Dozier did not laugh at his idea of talking to Mars. Both men were certain that the universe reached far beyond the city limits of Little London, or of this earth, or even this galaxy. It was a pleasant relationship.[6]

Although Tesla was intent upon starting his laboratory, he was to be plagued with a slight delay. Just three days after his dinner at the El Paso Club, it was impossible to get a workman on the job. The reason for this was that Little London had literally broken out in a rash of red, white, and blue and a frenzy of patriotic fervor for the visit of Admiral Winfield Scott Schley, the hero of Santiago Bay. Tesla chafed at the delay but he could not ignore the noisy hysteria, much of which centered in the Alta Vista Hotel. Who could sleep or even think while the hammers tore away the cigar stand in the lobby, and nailed red, white and blue bunting and the hideous Navajo rug (which had an American flag custom-woven into it), fast to the wall? Nor could Tesla find peace in a quiet walk along Little London's broad streets, for now they were filled with the holiday crowd, mounted police with their buttons polished and their horses curried and combed and bedecked with bridle-plumes of red, white and blue. There was no hope of evening bringing any more quiet, for a parade was timed for sunset, to be announced with a thirty-gun salute and the snappy Midland Band all decked out in Indian costume.

No, there would be no peace or quiet for the man in 207, until the reception for 3,000 guests in the downstairs parlor was over and the Admiral had departed clutching a bewildered young eagle (a gift of an imaginative young hunter in the hills) in his black and blue hand and a copy of Maude McFerran Price's "Rough Riders' March," in the other.[7]

Tesla was glad to have the hubbub die down at long last. He was anxious to look over the laboratory site again as early as possible. The morning air was always invigorating. He disliked crowds. He much preferred a few choice friends. He was proud of his American citizenship, but the trappings of war and its fanfare were anathema to him. His inventions had been created with the hope that they would enforce peace—yet he was aware that the world wanted to turn them into aggressive channels. In his thinking, wars were a waste, and the implements of war did not serve their purpose. Gunpowder had not brought peace. But there was a hurrah and stimulation of hypnotic frenzy to war that peace could never equal.

Tesla could hear young Julie stirring about in the room next to his. It was very late. Actually, Julie was too excited to sleep. She was looking at the signature which Admiral Schley had written in her little autograph book that evening. It was on the page next to the signature of Nikola Tesla.

On June 2, the *Evening Telegraph* related for the benefit of curious townspeople that Tesla's laboratory would soon be ready for the first of the apparatus to be installed. The report noted that

Tesla had become concerned over Colorado wind and had had the laboratory braced on three sides—"from which," wrote the reporter, "it may be inferred that Mr. Tesla is a great scientist but a poor architect." He also predicted that the civilized world would await the results of Tesla's Colorado Springs experiments.[8]

Although Tesla was absorbed in construction problems, he welcomed an occasional letter from his friend, Robert Underwood Johnson. Johnson's easy humor always relaxed and delighted him. "Now that you have got so high in the world," wrote Johnson, "I suppose you look down on your old friends, the Johnsons. All that we ask is that you send your first wireless message to us."[9]

Tesla's description in his notes of this photograph, made in late December 1899 by Dickson D. Alley, reads, "View of laboratory from the rear by moonlight; 1 h 20 min. exposure. Moon about 2/3 full. Showing Pike's Peak Range and all details of building very sharply. Such photographs by moonlight could be secured in only a few places. Plate same as before, 11"x14" isochromatic." *Courtesy Nikola Tesla Museum, Belgrade, Serbia, (006-04-20).*

As the equipment began to be installed, Tesla scheduled projects requiring great amounts of current, for evening hours when the city load would not be too heavy.

As usual, although he was confronted with problems of great magnitude, he was nonetheless always the perfectionist, being annoyed by the most trivial details. "Tell Mr. Uhlman" (an employee in New York), he once wrote to Scherff, "not to sign his letter, 'yours truly,' but rather to close with 'sincerely.'" Also he ordered writing paper sent from New York. Such petty details took on unwarranted size.

The New York laboratory had been rushing the materials for Tesla. Balloons of eight- and ten-foot dimensions had been bought from a balloon firm in Frankfort, and the crew was trying desperately to get them painted with some kind of varnish that would increase their durability. A Mr. Myers was advising them. At best, the balloons would scarcely lift a pound. Mr. Myers advised filling the balloons only two-thirds full at high altitude. Hydrogen was being considered.

Besides the balloons, three hundred bottles had to be packed and sent. Eleven hundred feet of solid wire went out in the first shipment. Oscillators and spools were being packed.[10]

There was still paper work to be done, and sometimes in the daytime Tesla would take his stool and go out into the incredible Colorado sunshine with his notebook and pencil. One little girl was a frequent visitor. She was curious about the weird building and the tall dark man. She would approach timidly but the strange man was always kind to her and would sometimes explain things to her although she could not understand. After a while she would wander off across the prairie, avoiding the patches of prickly pear with her bare feet, and looking curiously at the squatters who camped out on the plains.[11]

Sometimes Tesla saw the "city herd" as it was being driven out to pasture by Guy Parker and other herders. Many of the people in town owned their own cows and this was the practical way of pasturing them during the day. At night they were driven home before milking time, past the laboratory.

Occasionally a young visitor would be shunted off with a curt warning to get out and get going. Among the young boys who were excited about Tesla was young Richard Gregg, a boy of fifteen, who was the son of the Congregational minister. Young Gregg was half fearful to approach the inventor, so he asked Leonard Curtis to put in a good word for him before he went for an interview. He had no experience to offer—only a boy's insatiable desire to know. That with Curtis' recommendation got him the job.

Young Gregg went to work on a Monday morning. Tesla put him under the watchful eye of his engineer, Lowenstein. The genial red-bearded man was patient and kindly. Each day the boy swept up the place, wound the wire, and moved what was to be moved. He was a general handyman.

If Richard's family expected him to come home with glowing tales of the happenings at the laboratory, they were disappointed. Richard came home obviously exhausted and worried about his new work. When the family questioned him about his wages, he had to tell them that he did not know what Tesla would give him, but whatever it was, he hoped he might buy a bicycle to ride to work! He was so tired. One evening he came home particularly discouraged over his job. He had made an error and Tesla had called him "a little fool." When Saturday of his first work week arrived, Richard was called into Tesla's office. "How much do you think you are worth?" asked the inventor. The boy had no idea, for this was his first job. "Do you think that four dollars would be enough?" he asked the boy. Richard was tongue-tied and embarrassed and overcome at the thought of the princely sum of four dollars. When Tesla noted the boy's confusion, he gave him eight dollars in his envelope. Immediately the youngster became a jubilant creature. All the fatigue and anxiety of the past week rolled like magic from his shoulders. He was successful, reliable, and worth more than twice as much as he had thought! He raced home with the money and the news. What would he do with his fortune? His parents waited for his answer. With success, he did not need a bicycle—for he could run miles—the way the Book said—and not be weary. "I think," he told his mother, "I will save it for electrical books."[12]

Tesla fenced off the area around the building and posted signs with black letters on all the posts warning, "KEEP OUT—GREAT DANGER." The curiosity of news reporters sometimes led them to violate the warnings. With them, Tesla had no patience. He was dealing with such a concentration of power which could kill in an instant, and he showed no mercy to fools who insisted on rushing in. At times the very grass seemed to crackle with electricity and sparks an inch long could be drawn from the water hydrant at a distance of 300 feet from the laboratory.[13]

Hugh D. Harper, later to become Police Chief of Colorado Springs, was at that time a motorman on the street car that ran in that vicinity. He said the laboratory was an unforgettable sight but that at the time he wasn't interested enough to investigate. "I guess we all thought he was crazy," he commented.[14]

Occasionally, a few others who strolled past the laboratory noted that it was a weird contraption—a roof that could be rolled back—a wooden tower rearing eighty feet in the air. Beyond it soared a 200-foot mast with a huge copper ball that was obviously not for decoration. There were rumors by some who had sneaked under the ropes for a closer look that there were copper bars as big as two-by-fours. In the center of the room was a gigantic high frequency transformer and more than 40,000 feet of wire. A Tesla coil, the largest ever built, dominated the scene. The remaining space was filled with dynamos, electric wires, switches, generators, motors, and every conceivable gadget known to electricians. "Through this maze," said one window peeker, "Mr. Tesla walks as fearlessly as if on the streets of the city."[15] Although the Alta Vista Hotel was his mailing address, summer found him spending less and less time in his room. He ate breakfast at the hotel but usually took his lunch and had supper delivered to him.

Some mornings, pretty Julie Stevens would watch him as he left in the buckboard from the Antlers Livery, with his long legs hanging over the side. She knew that there was more reason for this posture than the fact that his legs were long and the buckboard short—she knew that he was afraid of the horse and she laughed.[16] Little Julie wasn't afraid to grab the reins of any horse she had ever seen. This tall dark man was very strange.

Notes

1. *Colorado Springs Evening Telegraph*, May 17, 1899.
2. Interview with Julie Hoefer.
3. Microfilm letters, Tesla to Scherff, May 21, 1899, Library of Congress.
4. *Mountain Sunshine*, July-August 1899, vol. 1, No. 1, 33-34.
5. Ibid.
6. Interview with Frank Waters.
7. *Mountain Sunshine*, 32.
8. *Colorado Springs Evening Telegraph*, June 2, 1899.

9. Microfilm letters, Johnson to Tesla, June 8, 1899, Library of Congress.

10. Microfilm letters, Scherff to Tesla, July 21, 1899, Library of Congress.

11. Interview.

12. Correspondence with Richard Gregg and Gregg family letters.

13. Richard Grove, "The Wizard of East Pikes Peak," *Colorado Magazine*, October 1958, Vol. XXXV, No. 4, p. 269.

14. Inez Hunt and Wanetta Draper, *To Colorado's Restless Ghosts*, (Denver, Colo.: Sage Books, 1960), 188.

15. *Colorado Springs Evening Telegraph*, June 21, 1899.

16. Interview with Julie Hoefer.

Tesla's Colorado Springs laboratory was built rapidly by local contractor and builder Joseph Dozier in late May and early June 1899. First experiments were conducted in the lab on June 15. The building is shown here in an early phase of construction, probably during the summer, and the entrance is cordoned off with signs saying "Great Danger—Keep Out." The building measured 50 by 60 feet and was located on a high, open plain on the north side of Pikes Peak Avenue, between the School for the Deaf and Blind and the Union Printers Home. The site was across from present-day Memorial Park, in the area of Foote Avenue. *Courtesy Nikola Tesla Museum, Belgrade, Serbia, (006-04-006).*

Various coils "differently attuned" in Tesla's Colorado Springs laboratory resonate with vibrations sent to them from an electrical oscillator. Tesla wrote in his Colorado Springs notes that about 100 separate flashes, or "throws of switch," produced the electrical streamers in this photograph, while the room itself was lighted with an arc lamp during a separate 15-minute exposure. *Courtesy Nikola Tesla Museum, Belgrade, Serbia, (006-02-071).*

Tesla Discovers Stationary Waves in the Earth

Tesla was increasingly content in his new laboratory and felt his choice of Colorado was a fortunate one. He set to work eagerly, filled with hope and anticipation. He was not oblivious to the beauty of the western skies or the imposing sight of high mountain ranges. The quiet and restfulness of the place made ideal conditions for his scientific observations. The exhilarating climate sharpened his senses and his feeling of physical well-being. He was sure that his vision was improved and his hearing, always acute, was even more keen. He was hesitant to account how far objects could be distinguished in the clear atmosphere, for fear of ridicule. He did venture to vouch that he had heard claps of thunder seven and eight hundred kilometers away. He declared:

> No better opportunities for such observations as I intended to make could be found anywhere. Colorado is a country famous for the natural displays of electric force. In that dry and rarefied atmosphere the sun's rays beat the objects with fierce intensity. . . . Aided by the dryness and rarefaction of the air, the water evaporates as in a boiler, and static electricity is developed in abundance. Lightning discharges are, accordingly, very frequent and sometimes of inconceivable violence. On one occasion approximately twelve thousands discharges occurred in two hours, and all in a radius of certainly less than fifty kilometers from the laboratory. Many of them resembled gigantic trees of fire with the trunks up or down. I never saw fire balls, but as a compensation for my disappointment I succeeded later in determining the mode of their formation and producing them artificially.[1]

Although the scientist found the little community quiet and restful, he couldn't have missed entirely the celebrations planned for Independence Day. He was engulfed in the wave of excitement over the lavish pyrotechnical display which was planned for the night of July 3 on the summit of Pikes Peak. The streets were already draped with flags and bunting. Youngsters and oldsters alike had laid in a supply of firecrackers to greet the dawn, but the highlight of the two-day celebration was to be the illumination of the area's famous mountain top. Such a feat had been attempted once before when some eager citizen took fifty barrels of tar to the top, but the total effect only amounted to a "ten-cent" bonfire. This time the officials had gone all-out and had transported thirty-five thousand pound boxes of powder to the summit and were sure that the red, white and blue flares would light the sky so far that the colors would be visible as far away as Cheyenne, Wyoming.

But nature conspired to interfere with an unscheduled show of her own. At sundown, Tesla, as well as the townspeople, noted a strange light which lingered in the sky far past the usual time, a brightness in the heavens which carried an atmosphere of augury, the kind of celestial light which is sometimes reported as coincidental with events of import. A local newspaper said:

> The brightness of the heavens that evening was certainly very great and the scene was one of exceeding grandeur. Most of it, however, was due to the elements, which seemed determined to play a prominent part in the proposed illumination.[2]

Tesla observed the dense mass of strongly charged clouds which gathered in the west, but he was concentrating his whole being on putting the elements to his own use and could not share the chagrin of the people in the streets of Little London when the pyrotechnical display lost in its competition with nature.

A reporter standing by on the summit of the great mountain was one of the few who considered the fireworks display worthy of the effort. He waxed poetic in his description of the scene from his high vantage point and declared that the clouds below added to the drama taking place on top of the mountain.

> At eight o'clock the first flare had been lit, but rain completely enveloped the area. At nine o'clock the first signal light was seen faintly as a match from below.

Then, he continued:

> The gods of the storm burst forth with one of the grandest electrical displays ever witnessed and for an hour the heavens were lit with living fire. The thunder rolled and as the storm moved westward toward Cheyenne Mountain a thousand fiery thunder-bolts reverberated through the canyons and rolled over the plains. The storm lessened slightly within an hour. The curtain of clouds parted.[3]

The storm deposited a two-inch blanket of sleet, hail and snow on the summit of Pikes Peak. The temperature dropped to twenty-eight degrees. The mountain stood like an ermine-crowned potentate, wrapped in the flames of the national colors as the powder was lit to outline the entire peak. For the final display the powder at all stations was ignited at once and Schley Rock, where the Admiral had been photographed early that summer, was lighted in the line of the illumination. The reporter enthused further: "Below, the lights of a half-dozen cities twinkled like brilliants, but in five directions the lightning zig-zagged in five separate areas." The national colors hung suspended in the turbulent clouds to form a transcendently beautiful, but eerie, spectacle. The summit house appeared to hang suspended like a phantom airship. The moment was dramatic, but brief. Almost immediately darkness again took over the scene and the remains of the spectacle were ashes on the snow.

With few exceptions, such as the ecstatic reporter and the New York inventor, there was no such enthusiasm among the people of the little town at the foot of the Peak. The eager souls who had ventured the storm had waited under their umbrella observatories. As the clouds had parted, the white outline of the Peak was sharp against the darkness, but for all the to-do about the gigantic display, it really hadn't amounted to much. Some were openly critical of the puny show and petulant over the waste of effort and money. They resented the elements which had dared to overturn their plans, but some of them observed, "It always rains for Fourth of July," and scheduling the celebration a day early had not held off the elements.

Nikola Tesla felt no such disappointment. The storm was made to order for him. He had been watching it from the beginning and was keenly aware of the strange brightness of the heavens with their atmosphere of augury. For him, it was the eve of what he considered his greatest discovery, the threshold of the first decisive experimental evidence of a truth of overwhelming importance for the advancement of humanity. He had observed closely the dense mass of strangely charged clouds that had gathered in the west in the early evening of that July 3, and the violent storm which followed filled him with excitement. As he watched the moving clouds, he began to observe a pattern to their formation and later said:

> . . . a violent storm broke loose after spending much of its fury in the mountains. It was driven away with great velocity over the plains. Heavy and long persisting arcs formed almost in regular time intervals. My observations were now greatly facilitated and rendered more accurate by the experiences already gained. I was able to handle my instruments quickly and I was prepared. The recording apparatus being properly adjusted, its indications became fainter and fainter with the increasing distance of the storm until they ceased altogether. I was watching in eager expectation. Surely enough the indications began again, grew stronger and stronger, and, after passing through a maximum gradually decreased and ceased once more. Many times, in regularly recurring intervals, the same actions were repeated until the storm which, as evident in simple computations, was moving with nearly constant speed,

> had retreated to a distance of about 300 kilometers. Nor did these strange actions stop then, but continued to manifest themselves with undiminished force. Subsequently, similar observations were also made by my assistant, Mr. Fritz Lowenstein, and shortly afterward several admirable opportunities presented themselves which brought out still more forcibly and unmistakably, the true nature of the wonderful phenomenon. No doubt whatever remained; I was observing stationary waves.[4]

The earth was proving itself a conductor, and the tremendous significance of this fact in the transmission of energy became clear to Tesla. He was convinced that not only could he send telegraphic messages to any distance without wires, a fact he had long before recognized, but also that it would be possible to impress upon the entire globe the faint modulations of the human voice, and even more wonderful, to transmit unlimited power to any terrestrial distance with scarcely any loss.[5]

The following year, in his famous article in *Century Magazine,* he expanded on the great significance of this discovery.

> Stationary waves in the earth mean something more than mere telegraphy without wires to any distance. They will enable us to attain many important specific results impossible otherwise. For instance, by their use we may produce at will, from a sending-station, an electrical effect in any particular region of the globe; we may determine the relative position or course of a moving object such as a vessel at sea, the distance traversed by the same, or its speed; or we may send over the earth a wave of electricity traveling at any rate we desire, from the pace of a turtle up to lightning speed.[6]

The morning of Independence Day dawned with Tesla feeling a wonderful sense of elation and great accomplishment. He was aware of the timbre of the small city. The war fervor had not died down. War bred heroes; science bred martyrs. No one was in the mood to listen to the discoveries of a lone and eccentric scientist.

The holiday issues of the local papers whined through the little troubles—the disappointments of the night before, the rain, the weary policemen, the firecrackers which had blown up a hitching post, and on the national scene, Cuba, which was presenting a few problems.

But the morning had brought a sense of renewed anticipation to the little town. After all, there were still fireworks to be set off. The fire department stood by to cope with a rash of flames, but the brave lads were not called out even once. Only one small incident marred the celebration and it was of import only to the one person mainly concerned. Too much enthusiasm had been vented on the Chinese laundry and firecrackers had blown it up, but this was a trifling prank which the Chinese laundryman should have overlooked as merely an excess of American enthusiasm for their national holiday.[7]

The papers were still complaining about the feeble illumination on the Peak on the previous night. Cheyenne had reported seeing a feeble spark from the area, as did Denver and Pueblo. Cripple Creek wasn't much concerned with Colorado Springs' displays, anyway. They were busy getting ready for a bigger and better one of their own on their signal mountain, Mt. Pisgah. This was planned as the climax of their tremendous three-day celebration when their own mountain was to be turned into a veritable volcano by means of 150 sticks of dynamite and fifty cords of kerosene saturated wood, so they naturally had little interest in the feeble efforts of Colorado Springs in the celebration of the glorious Fourth.[8]

Tesla made another discovery that summer which he considered of cosmic importance. Alone in his laboratory one night he became aware of signals coming in a pattern. He became increasingly confident that these were definitely an attempt at interplanetary communication from either Venus or Mars. He admitted to being terrified in that moment of revelation but was certain that he would find a way to return the signals. He was positive that it was a definite attempt at communication, not of terrestrial origin, nor atmospheric, nor influenced by the Aurora Borealis, but that it was a message with a distinct one-two-three pattern—the first ever to have been received.[9] To report his theory at that time would have been detrimental to his proven work, but his new theory was of such magnitude that he could scarcely contain his excitement and awe. There was only old Mr. Dozier with whom he could share his discovery. Other Little Londoners would scoff or be politely tolerant, he was certain. Actually, when his discovery was reported in the local newspapers at a later date, they were more kindly than some of his colleagues in the scientific world.

A Colorado Springs newspaper said:

> If there are people in Mars, they certainly showed most excellent taste in choosing Colorado Springs as the particular point on the earth's surface with which to open communication. In fact, we may feel assured that if the mystical one-two-three, which Tesla says may have been impulsed from Mars, should be translated from the planetary code it would read, "How is the weather in Colorado Springs?" or something equally localized. Unfortunately other cities and other scientists do not feel this same consideration towards Mr. Tesla. Professor Holden, for instance, formerly Director of Lick Observatory, in a current magazine says, "Mr. Nikola Tesla has announced that he is confident that certain disturbances of his apparatus are electrical signals received from a source beyond the earth. They do not come from the sun, he says; hence they must be of planetary origin, he thinks; probably from Mars, he guesses. It is the rule of a sound philosophizing to examine all probable causes for an unexplained phenomenon before invoking improbable ones. Every experimenter will say that it is almost certain that Mr. Tesla has made an error and the disturbances in question come from currents in our air or in the earth. How can anyone possibly know that unexplained currents do not come from the sun? The physics of the sun is all but unknown as yet. At any rate, why call the currents 'planetary' if one is not quite certain? Why fasten the disturbances of Mr. Tesla's instrument on Mars? Are there no comets that will serve the purpose? May not the instruments have been disturbed by the great

Bear or the Milky Way or the Zodiacal light? There is always a possibility that great discoveries in Mars and elsewhere are at hand. The triumph of the scientists of the past century are still striking proof, but there is always a strong probability that new phenomena are explicable by old laws. Until Mr. Tesla has shown his apparatus to other experimenters and convinced them as well as himself, it may safely be taken for granted that his signals do not come from Mars."[10]

The newspaper was unwilling to accept Professor Holden's criticism and countered:

As a scientific guesser Prof. Holden is not up to the Colorado Springs altitude. Mr. Holden believes in little or no water in Mars, a temperature far below the freezing point, little or no air, polar caps composed of frozen carbon dioxide, and "canals" that are in reality long tracks or crevasses in the planet's outer crust.

Frankly, we prefer the Tesla guess to that of Holden. It is a good rule in inventional science when you're going to tell one, tell a good one, and men have become great by observing the rule.

The man in the moon is gone and even H. G. Wells cannot bring him back, but we refuse to give up Mars. If the selection of Colorado Springs as a point of superior knowledge where Martian signals would be most likely to be noticed and understood is not a sign of superior intelligence, what more could be expected? For the present we stand by Mr. Tesla, and we have no doubt that ere long he will not only catch the repetition of his one-two-three message, but that he will be able to announce to the Martians the quotations and sales of the Colorado Springs Mining Stock Exchange.[11]

It is quite possible that the signals which Tesla received were radio waves coming from the stars. It was not until 1932 that scientists accidentally discovered that the source of some static was from outer space and they began to study the signals as they came and the signals were transmitted in the form of coded numbers to be fed into a digital recorder.[12] Now, just for the enjoyment of it, radio astronomers occasionally hook up their antennae and amplifiers to loud speakers and listen to the hissing of the Milky Way, the sighing of the sun, and the mournful grumblings of Jupiter.

In I960, Dr. Frank Drake of the National Radio Observatory began listening for definite signals from other worlds. For ten minutes after he tuned in his second star, Tau Ceti, he observed a clear beep-beep from some unexplained source. Some scientists feel that the world's last, best hope may lie in its contact with extra-terrestrial brains whereby man may exchange technological knowledge.[13]

The world is less inclined to scoff at such "far-out" theories in the 1960s than it was in 1899.

The circle of Tesla's friends in Colorado was small and select; the number of those to whom he could advance his wildest theories was still smaller. He enjoyed his evenings at the home of E. G. Eaton, a Colorado Springs banker, and the conversation of their friends. It was at the Eaton's home that he told, one evening, that he had illuminated lights the distance of half a mile from his laboratory by the use of earth-conducted power. Young Leo Bortree, a pre-medical student at Cutler Academy, listened to Tesla that evening and found him charming, but certainly *avant-guard.* Nevertheless, Bortree was curious enough to go past the weird laboratory. It was a sight he would never forget, even years later when it was reduced to a memory and the concrete foundations had crumbled.[14]

There was a time when Tesla had observed lightning and had predicted the exact moment a concussion would reach the laboratory from a striking point ten miles distant. But this was another radical thought, beyond the interest or ken of most of his new friends. How could he explain to them that his hearing was so phenomenally keen that it wasn't remarkable for him to detect thunder-claps at a distance of 550 miles? Even his young associates, so carefully chosen, could scarcely note thunder more than 150 miles.[15]

The community had expected immediate concrete evidence of Tesla's genius, but gradually came to look upon his laboratory goings-on with tongue in cheek. There were rumors that he set his building afire at times by pressing electric buttons to see how general a conflagration he could start and that he employed his own private fire department to extinguish the flames. There was more truth in the rumor that he had no practicality in money matters and that at one time, when he was unable to procure barrel hoops, he came to the city and ordered a wagon-load of bicycle rims for his purpose.[16]

His generosity to tradespeople was commented on. None of them went away without tips. Glen Jackson, a young lad working for a Colorado Springs jewelry firm, welcomed the opportunity to make a delivery to Tesla on the off-chance that he might get a peek into the laboratory. Tesla was in need of gold-filings and he had two twenty-dollar gold pieces made into filings for the order. Jackson delivered them personally, whereupon the inventor reached into his pocket and, without counting the change or noticing the coins, gave the boy a handful of French money, one coin of which he saved to treasure all his life.[17]

Friends in New York were equally aware of Tesla's lack of ability in money matters. Robert Underwood Johnson wrote jestingly, "Is it true that you are going to stay in Colorado and run for the Senate as a millionaire?" In the same letter, Johnson reminded Tesla that he, Johnson, had ten thousand dollars salted down in the Argentium-Juniata silver mine at Aspen, apparently in the hope that he also might become a millionaire.[18]

Inquisitive reporters were nagging Tesla for dramatic news stories, but they were getting less and less friendly welcome from him when they irritated him with their sensation seeking. One young man needing headline news for a July article asked, "When are you going to make an experiment in wireless telegraphy?"[19] Here, he revealed his ignorance of Tesla's late accomplishments. The reporter wondered if because of Marconi's achievement, Tesla could have felt that practical application of wireless was already a certainty. It was characteristic of him that he lived in the "theoretic stratosphere" and was eagerly pursuing his next idea. What the reporter could not realize was that Tesla was absorbed in a completely revolutionary theory about power transmission, and wireless communication.

So when the newsman peered into the laboratory at the large coil which could send out three million volts, and at the smaller one which could be used for lesser volume, and the twelve large storage batteries placed around the transmitter, and then asked Tesla when he would start making experiments, the scientist countered that he had been making them all the time, ever since his arrival in Colorado Springs, then added, impatiently, "I think that is all I wish to say," and disappeared into his private office in the laboratory.

Only one local photographer, Fred Stevens, was allowed to set up his camera in the prairie laboratory. When Robert Underwood Johnson became aware of the scope of his friend's Colorado Springs experiments, he began to plan for an article in *Century Magazine,* with accompanying photographs.

Notes

1. *Nikola Tesla, 1856-1945, Lectures, Patents and Articles,* (Beograd, Yugoslavia: Nikola Tesla Museum, 1956), A-153-154.
2. *Facts,* July 8, 1899, vol. iv, no. 40.
3. Ibid.
4. *Lectures, Patents and Articles,* A-156.
5. Ibid.
6. Ibid., A-149.
7. *Colorado Springs Evening Telegraph,* July 4, 1899.
8. Ibid.
9. Nikola Tesla, "Talking with Planets," *Current Literature,* March 1901, 359.
10. *Colorado Springs Gazette,* March 9, 1901.
11. Ibid.
12. David Bergaminia, *The Universe,* (New York, N.Y.: Time, Inc., 1962), 58.
13. Ibid., 38.
14. Interview with Dr. Leo W. Bortree.
15. Nikola Tesla, "My Inventions," *Electrical Experimenter,* April 1919, 907.
16. *Colorado Springs Gazette,* September 6, 1905.
17. Interview with W. Glen Jackson.
18. Microfilm letters, Johnson to Tesla, July 24, 1899, Library of Congress.
19. *Colorado Springs Evening Telegraph,* July 21, 1899.

Perhaps the most spectacular and well-known photograph of Nikola Tesla was made in his Colorado Springs laboratory, December 1899, by himself and Dickson D. Alley. The electrical discharges, running about twenty feet from the central coil to another vertical coil on a stand, were photographed in the completely darkened room with about fifty short "throws of the switch." The photographic plate was then exposed again, with Tesla seated in the chair, using Eastman flash powder for illumination. *Courtesy Nikola Tesla Museum, Belgrade, Serbia, (006-02-138).*

Tesla's Most Dramatic Colorado Experiment

There were a few more experiments to be made before Tesla closed his Colorado Springs laboratory and packed his bag to return home. He was almost reluctant to leave this invigorating country. He had always been vulnerable to respiratory infections but this climate had been good. September was intoxicating. Once after an early fall storm he wrote to Robert Underwood Johnson:

> I wish you could see the snowdrifts and icebergs of Colorado Springs. I mean those that float in the air. They are sublime. Next to your poems, Luka, the finest things on earth![1]

The townspeople had ceased to pay much attention to Tesla and his crazy laboratory. The community was exhausted from its tourist summer and the big Flower Parade. Half the carriages and bicycles in town had been covered with crepe paper flowers and streamers. It had consumed so much time and effort and inventive ability. Tesla did not notice any neglect, for the time was growing short, funds were growing low, and he was about to solve important problems for all mankind. He wrote to Scherff:

> The progress of my work here is very satisfactory.
> It was a good thing to come, all except financially.[2]

He was jubilant in his success with power transmission through the earth by means of his magnifying transmitter. Full records of his experiments will never be known, for many of his plans he carried in his retentive mind, not trusting his secrets on paper. He even required his technicians to be able to work without drawings. He could visualize every section fitted and every measurement worked out to perfection. He could detect if some machinery would be out of balance in his mental picture. Confidential letters revealed a meagre, optimistic report to Scherff. However, later, a written technical description was published concerning his magnifying transmitter. In this report he made the claim that while in Colorado, he had once set up two hundred incandescent lamps twenty-six miles from his laboratory and lighted them with power thrust through the earth. He used no wire connection to any generator. In this technical report, he claimed that this could be done anywhere around the globe and that distance would not deter the transmission with his equipment.[3]

Shortly after the September storm that he had described to Johnson, he sent word to Scherff to have Kolman Czito, from the New York laboratory staff, join him for the most dramatic experiment of his career. Scherff knew that the big moment was drawing near as he telegraphed to Tesla that Czito would leave New York on September 20th.[4] He followed the telegram with a letter saying,

> Mr. Lowenstein has told me and Uhlman something of your wonderful work there and we know that instead of a century you are a thousand years ahead of the others.[5]

Tesla ignored the compliment. He was intent now on delivery of much needed equipment from New York. He was so engrossed in imminent experiments that he did not comment when Scherff informed him that Lee De Forest (an earnest disciple, destined for radio fame) had applied for work in the New York laboratory.[6]

Again it was the fever of recent war with Spain and the firecracker hysteria of war that was to delay Tesla. Not Schley, this time, but George Dewey in New York, played havoc with the work schedule while New Yorkers went wild with two Dewey holidays. Scherff had his hands full.

The exhilaration which Tesla had felt in early September began to flag. A reporter from the *Evening Telegraph* questioned him about his health and Tesla admitted the grueling hours he was working were taking toll. "In New York," he explained, "I work until two or three o'clock every morning. I cannot change my working habits." Tesla discussed Marconi with the reporter. Then Tesla said, "Nikola Tesla is about to be heard from again in a way that will interest the entire world!" The newsman wound up his report with the words,

> There is something weird and yet nothing unreal about this man whose happiest moments are during an electric storm when the wires snap about him and light up on all sides, when it would seem his life is in danger. No man will doubt a man like this when he says, "I will be heard from."[7]

Tesla appears to read calmly while symmetrical electrical discharges from an eight-inch ball on the central coil dance around him. For technical reasons, the image of the electrical discharges and the image of Tesla in the chair were made using separate exposures. *Courtesy Nikola Tesla Museum, Belgrade, Serbia, (006-02-142).*

It was October 11th before some badly needed batteries could be boxed and shipped to Colorado Springs.

At last came the time for the experiment which was to mark the high hill of Tesla's career. For some time he had been working with the apparatus which produced lightning-like discharges. With the approaching time for the experiment, he inspected every piece of equipment thoroughly. He intended to produce far higher voltages than had been produced in the high-voltage transmission lines at Niagara Falls.[8] He was confident he could duplicate celestial lighting by man-made means. In his mind he knew just how the tremendous currents would behave, yet he was aware of the chance of unpredictable disaster. With confidence that was not arrogance, he affirmed his belief in himself. "I make no mistakes. I never fail!" He realized that he would be dealing with apparatus which could kill 30,000 to 300,000 in one instant, yet the drama he was about to evoke demanded a close front seat for himself and Kolman Czito, his chosen assistant.

The interior of Tesla's Colorado Springs laboratory, showing the outside of the circular oscillator frame, at left, and a bank of condensers (capacitors) in the foreground. *Courtesy Nikola Tesla Museum, Belgrade, Serbia, (006-04-024).*

Everything was finally in order. He planned that the flames would strike from the copper ball at the top of the mast and he hoped that his calculations were right. Tesla and Czito were at last alone and ready for the zero hour at hand.

Tesla warned Czito, "We will close the switch only for a second and then quickly open it." When they made sure of the operation, Tesla would stand outside the laboratory to view the results.

Czito noted that Tesla was a gaunt Mephistopheles that night, wearing full evening attire, impeccable except for boots which made him an inch taller because of thick insulator soles. When Tesla dressed in this manner it was indication of the importance of the moment. Czito was aware of the danger they were facing, but he considered it an honor to be chosen for this night's work. He hoped to tell his grandchildren all about it someday—if he lived to tell it!

Electrical streamers discharge symmetrically from the central coil in Tesla's Colorado Springs Laboratory. *Courtesy Nikola Tesla Museum, Belgrade, Serbia, (006-02-089).*

There was a moment of suspense as Czito rested his hand on the switch. In his own mind he believed that even a brief contact would short-circuit the whole thing, yet Tesla had a way of working things out. He was painstakingly deliberate. After an interminable time, Tesla called out, "Now! Czito, close the switch." Czito obeyed and jerked it open again according to direction. There was no short circuit. Instead, a Medusa-headed fire glowed around the secondary and there was a crackling sound and a sharp report from the ball above. Another tryout convinced them they were ready for the big scene.

"Now," said Tesla, "I must stand outside where I can see the top of the mast perfectly. I will give the signal and you are to leave the switch closed until I signal you to open it." A few minutes later he called, "Czito, close the switch." Czito obeyed orders and then jumped back to a distance from which he could yet reach the switch on command. Now he was certain the short circuit would come. The crackling and snap repeated and then came a tremendous upsurge of sound as the power built up. There was a crescendo of vicious snaps from above. The noises became machine-gun staccato—then roared to artillery intensity. Ghostly sparks danced a macabre routine all over the laboratory. There was a smell of sulphur that might be coming from hell itself. A weird blue light spread over the room. Flames began to jump from the ball at the top of the mast—first a few feet long—then longer and brighter—thicker, bluer. More emanations until they reached rod-like proportions, thick as an arm and with a length of over a hundred and thirty feet. Tesla estimated the length by comparing the flashes with the laboratory. The heavens reverberated with a terrific thunder that could be heard fifteen miles over the ridge in Cripple Creek.[9]

Suddenly the inferno ceased with dramatic silence. Tesla screamed a rebuke to Czito for opening the switch. But Czito had not touched the switch. There had been a power failure. Tesla made a frantic call to the electric company, pleading with them not to interfere with his experiments. His call brought only an indignant reply from the company employee. The boys on the night shift were having their hands full with a blazing generator. They informed him that he had plunged the town into darkness and ruined their generator and that this would be the last of his free power from the Colorado Springs Electric Company, Mr. Curtis or no Mr. Curtis. From then on, he was told he would have an independent dynamo and it would be the one he had burned up—*if* and *when* he could repair it![10]

Tesla surveyed the silent laboratory where only moments before he had stood on the crest of a hill, at the apex of his life. The heavens had opened with his key and revealed a split-second of glory, and a glimpse of the future he could predict for the world. The Force which he had always pursued had on this night made obeisance to him.

Colorado Springs Electric Company power plant, ca. 1899. According to Ellison Hawks in *Pioneers of Wireless*, the Electric Company "was alarmed to find that the armatures of their big dynamos were suddenly burnt out, or fused, for no apparent reason. . . . the trouble was traced to the experiments that Tesla was conducting in his laboratory." One of Nikola Tesla's experiments reportedly resulted in a town-wide power outage. *Fred P. Stevens photograph. From Special Collections, Pikes Peak Library District, (005-4515).*

It must have been a reverent moment, but it is doubtful if he gave any thanks to his Maker, for Tesla had ceased to believe in a Supreme Being as an entity. At least not in the limited conception held by so many of his time. There was no scoffing in his unbelief. He was, simply, one of God's earnest doubters; one who would have commanded the respect of an unrecognized God.

On this night, the riddle of life seemed very close to his grasp. Revelation had been granted to him, whose altar offering had been the complete dedication of his life. For Nikola Tesla there had never been a ram in the thicket.

Reward enough for such sacrifice was this moment of illumination when poetry, mathematics, music, and science had fused by a new alchemy to become a key to the universe.

On this night he had wrested the power of heaven from the sky. He had stood for a fleeting moment with lightning in his hand.

Notes

1. Microfilm letters, Johnson to Tesla, September 16, 1899, Library of Congress.
2. Ibid., Tesla to Scherff, September 12, 1899, Library of Congress.
3. Nikola Tesla, "My Inventions," *Electrical Experimenter*, June 1916, 177.
4. Microfilm letters, Scherff to Tesla (telegram), September 18, 1899, Library of Congress.
5. Ibid., September 29, 1899.
6. Ibid., September 26, 1899.
7. *Colorado Springs Evening Telegraph*, "Nikola Tesla Talks of Marconi's Work," October 7, 1899.
8. John J. O'Neill, *Prodigal Genius*, (New York, N.Y.: Ives Washburn, Inc., 1944), 183.
9. Ibid., 186.
10. Ibid., 187.

Description of the experiment taken from John J. O'Neill's Prodigal Genius, *and Tesla's own description in* Century Magazine, *June 1900. [The vivid details of the often repeated story about the Colorado Springs experiment are not entirely evident in primary sources].*

"Electric Light Station," Colorado Springs, ca. 1899. According to an inteview with Nikola Tesla published in *The Electrical Experimenter*, August 1917, Tesla said, "As an example of what has been done with several hundred kilowatts of high frequency energy liberated, it was found that the dynamos in a power house six miles away were repeatedly burned out, due to the powerful high frequency currents set up in them, and which caused heavy sparks to jump thru the windings and destroy the insulation! The lightning arresters in the power house showed a stream of blue-white sparks passing between the metal plates to the earth connection." *From Special Collections, Pikes Peak Library District, (001-4791).*

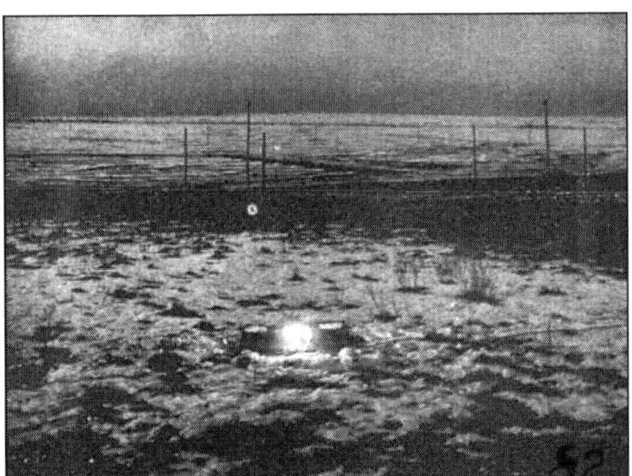

Three lamps are lit in a field near Tesla's Colorado Springs laboratory in a demonstration of the wireless transmission of energy. *Courtesy Nikola Tesla Museum, Belgrade, Serbia, (006-C-082).*

New York, January 1, 1904

I wish to announce that in connection with the commercial introduction of my inventions I shall render professional services in the general capacity of consulting electrician and engineer.

The near future, I expect with confidence, will be a witness of revolutionary departures in the production, transformation and transmission of energy, transportation, lighting, manufacture of chemical compounds, telegraphy, telephony and other arts and industries.

In my opinion, these advances are certain to follow from the universal adoption of high-potential and high-frequency currents and novel regenerative processes of refrigeration to very low temperatures.

Much of the old apparatus will have to be improved, and much of the new developed, and I believe that while furthering my own inventions, I shall be more helpful in this evolution by placing at the disposal of others the knowledge and experience I have gained.

Special attention will be given by me to the solution of problems requiring both expert information and inventive resource—work coming within the sphere of my constant training and predilection.

I shall undertake the experimental investigation and perfection of ideas, methods and appliances, the devising of useful expedients and, in particular, the design and construction of machinery for the attainment of desired results.

Any task submitted to and accepted by me, will be carried out thoroughly and conscientiously.

Laboratory, Long Island, N. Y.
Residence, Waldorf, New York City.

Nikola Tesla

"A striking Tesla Manifeso" featuring Wardenclyffe Tower. The first of four pages was published in the February 6, 1904, issue of *Electrical World and Engineer* stating,"We reproduce herewith in slightly reduced fac-simile the first page of a four-page circular which has been issued this week by Mr. Nikola Tesla in a large square envelope bearing a large red wax seal with the initials, 'N.T.' . . . The circular is an extremely interesting one. It is most sumptuously got up on vellum paper and altogether constitutes a manifesto worthy of the original genius issuing it. It is to be gathered from the circular that Mr. Tesla proposes to enter the field of consulting engineership, in which he already has enjoyed an extensive connection here and abroad."

Tesla's Return to New York: The Construction of Wardenclyffe Tower

Tesla arrived in New York on January 21, 1900. Money was, of necessity, occupying a large place in his thoughts. The little luxuries, seemingly indispensible to him, were also essential to his efficiency. He felt no humility in asking for financial assistance, for he had made millionaires for industry and kept nothing for himself. Every dollar he possessed was manna to be used, not hoarded. He was a spender, but not a spend-thrift.

It was naturally to Westinghouse that he first appealed upon his return. The very day after his arrival in New York Tesla wrote to his former employer:

> Please receive the following communication as a personal one. I have just returned from Colorado where I have been carrying on some experiments since a few months past. The success has been even greater than I anticipated, and among other things I have absolutely demonstrated the practicability of the establishment of telegraphic communication to any point of the globe by the help of the machinery I have perfected. In carrying out the plan I shall want a direct connected engine and dynamo of at least 300 horse power, but preferably more, on either side of the Atlantic, and as this would involve a considerable expense which in view of the apparent impossibility of the problem, and also because of panicky feeling all around, I fear I would have difficulty in securing, I wish to ask you whether you would not meet me on some fair terms in furnishing me the machinery, retaining the ownership of the same and interesting yourself to a certain extent in a way which I cannot yet clearly specify, but which might be found after a mutual exchange of ideas on the subject. I am naturally prompted by self interest in making this

> proposal, but you may believe me, that there is also a sincere desire to advance your own. The demonstrations which I have made in Colorado are of such a nature that they preclude the possibility of a failure, and the performances of the machines which I have developed are of such a character as to almost surpass belief. . . . I have been so enthused over the result achieved and have worked with such passion, that I have neglected to make such provisions for money as would have been dictated by prudence and on my return I find that I may have to make some payments before certain funds which are due to me, may reach me. Being thus compelled to borrow money I turn to you to ask whether your company will not advance me say $6,000 on the guarantee of my royalty rights from England, or if preferable, whether they would not buy outright my claims on the royalty for a sum of $10,000. I would prefer to borrow money and pay the interest as I am not quite sure that I ought to sell my share, which is four-ninths of the total amount due to my company, not knowing exactly, whether, in so doing, I might hurt the interests of my associates, Mr. Brown and Mr. Peck. Needing, however, the money and thinking that my right in your hands, would, if anything, enhance the value of their share, since it would give your company an additional cause to push the business, I have ventured to make this suggestion.[1]

The Westinghouse Company had all the facilities required at its command and Tesla felt no hesitancy in calling upon them for assistance. Had he not torn up a contract at a time when Westinghouse was in trouble—a contract which might have netted him a fortune?

There were others to whom he could appeal. At this point in his career, financiers came to him to learn the results of the Colorado experiments. Among the curious were such men as Astor, Morgan, and Ryan, all eager to learn what could be wrung out of the latest Tesla achievements. Of first magnitude to the inventor was his Colorado success in the transmission of power without wires. With his head still in the clouds, he told the financial magnates of his accomplishments and assured them that he could put power into the earth and command it to do his bidding. He could send heat to the North Pole and ice to the tropics, pictures around the world, music through the air, and time the clocks of the universe. The process could be worked out so that it was inexpensive and men could draw electricity unlimited from the earth at will. Men's labor would be lessened; peace and prosperity would be assured.

It was not peace nor ease for the little man which concerned the financiers. Even if all this were true, how could one meter such power and collect the bills? Hard-headed businessmen knew where capital came from. The little man was not their concern. John Pierpont Morgan, perhaps more than others, saw the scope of Tesla's ideas, but recognized the futility of a Utopian scheme in a utilitarian world. The financiers turned thumbs down on Tesla's ideas. Morgan alone remained somewhat interested in the inventor. "I could build a tower," said Tesla, "and with my system could produce a world broadcasting system that could earn

John Pierpont Morgan, ca. 1902. Morgan was a banker, railroad financier, and founder of U.S. Steel Corporation. He backed Nikola Tesla's Wardenclyffe Tower construction with $150,000, however, he refused to provide Tesla additional needed funds to complete the project. *Library of Congress Prints and Photographs Division Washington, D.C., (LC-USZ62-94188).*

millions for inventors." Morgan pricked up his ears. Tesla felt no humiliation in asking for money for he had the firm conviction that all knowledge belonged to mankind and that wealth should be freely given to disseminate this knowledge. He believed in *noblesse oblige.*

As for Morgan, money was not for extending knowledge but for accruing and compounding interest and extending investments. The House of Morgan was not founded on wild speculation.

Morgan had a genuine admiration for Tesla and confidence in his ability. He decided to back Tesla in his proposition. It has been stated that Morgan simply gave Tesla $150,000 with no strings attached. Actually, there were plenty of strings attached. Morgan delayed his check for a few months. Finally it came with the stipulation that fifty-one percent of the patents relating to wireless telephony and telegraphy, not only those to be used in the present but the ones to be developed—all were to be in Morgan's name. The $150,000 was well secured.

Tesla was willing to make any arrangements to get started. He wrote Morgan:

> . . . permit me to remind you that had there been only fainthearted and close-fisted people in the world, nothing great would ever have been done. Raphael could not have created his marvels, Columbus could not have discovered America, the Atlantic cable could not have been laid. You, of all, should be the man to embark boldly in this enterprise.
>
> . . . The control is yours, the larger part is yours. As to my interest—you know the value of discoveries and artistic creations—your terms are mine.[2]

On March 1, 1901, Tesla sent to Morgan his contract, signing over the fifty-one percent interest in his patents and inventions and in any future ones relating to electric lighting and wireless telegraphy or telephony.[3]

Tesla felt some disappointment that the other men on whom he had relied did not rush eagerly to his assistance, yet he knew their limitations and could wait. Besides, he had work to do and $150,000 was a start. With his enthusiasm and confidence at full peak he wrote to Morgan:

> What I contemplate and what I can certainly accomplish, Mr. Morgan, is not a simple transmission of messages without wires to great distances but it is the transformation of the entire globe into a sentient being as it were, which can feel in all its parts and through which thought may be flashed as through a brain . . . From one single plant thousands of trillions of instruments could be operated, each costing no more than a few dollars, and situated in all parts of the globe.[4]

Robert Underwood Johnson, editor of *Century Magazine*, was eager to share Tesla's Colorado experiments with the intelligentsia who read his publication. As both editor and friend he wanted the world to share the achievements which Tesla had told him had been brought about in that weird laboratory at the foot of Pikes Peak. To that end, Johnson asked Tesla to submit an article—a popular account designed for the intelligent layman. To his consternation, when Tesla submitted his manuscript, it contained little of the news for which Johnson felt the world waited. He complained:

> You are giving people Euclid and they don't want Euclid. They will say it is obscure and dull when it is only deep. The ordinary brain reels in following you through the giddy mazes of space.[5]

The article was revised time and time again, but finally Johnson gave up and printed it as Tesla wrote—a philosophical treatise on "The Problem of Increasing Human Energy." In it, Tesla contemplated human life, asking whence does it come, what is it, whither does it tend? He maintained that "the sun is the past, the earth is the present, the moon is the future" and further declared "the earth will be destroyed not by fire, but by ice," but mankind need not despair. He gave as his reasons for hope his firm conviction:

> There will still be left on earth a glimmering spark of life and there will be a chance to kindle a new fire on a distant star . . . Birth, growth, old age and death of an individual, family, race or nation, what is it all but a rhythm.[6]

He spoke of the universal laws which governed all, and tried to explain his growing conception of man as more than a mere automaton by reason of the creative principle of life; but, lapsing once more into the role of scientist, he reduced the problem to a kind of mathematical formula in which man and his opposing negative forces did constant battle—and for which he offered a mathematical solution. His manuscript rambled to cover such controversial questions as the use of whiskey, wine, and other stimulants, and he suggested that man could be better concerned with impure water which caused more deaths than whiskey. Recalling his hard-earned education in youthful gambling, he condemned it, too, as a negative factor. And meat-eating he classified as a barbarous habit; furthermore, the wanton and cruel destruction of animals was inimical to the morals of man. He insisted that we must free ourselves from the animal instincts which come from eating meat.

Tesla philosophized concerning man's inadequate conception of existence, declaring that even a crystal was a form of life. He carried the idea to dizzier heights when he said that there might be intelligent beings on other planets which, because of varying conditions, could subsist in a form unknown to us. Then, in a statement wholly unintelligible in 1900, but not so easily scorned sixty years later, he declared:

> We cannot even with positive assurance assert that some of them might not be present here, in this our world, in the very midst of us, for their constitution and life manifestation may be such that we are unable to perceive them.[7]

Among the retarding forces of the world he took his stand with Buddha—that "Ignorance is the greatest evil in the world" — an evil which he identified with the power of friction working toward destruction.

Included in negative forces was war, but Tesla predicted that the flying machine could bring on universal peace. He had lost a bit of his optimism, however, for he said, "for all that, men will fight on merrily."

Finally, he did get around to describing some of his Colorado Springs experiments and the published article was accompanied by pictures of the huge coils and oscillators, his magnifying transmitter, and some of the lights without wires.

Tesla went on to predict the use of solar energy, the possible application of terrestrial heat, and the coming age of aluminum. He discussed the potentials of stationary waves which he was convinced he had discovered in Colorado and by which, he said, we could completely control electrical space and force for all kinds of power transmission. He was certain that interplanetary communication was quite probable, if, indeed, he had not already experienced it.[8]

Johnson editorialized about the article:

> Tesla presents a new and never to be forgotten conception of humanity moving as a mass. To follow his vivid introduction is like looking down from a great height and seeing familiar and new regions in their true relationships. It begets the feeling of being a spectator at the progress of the world, of listening to the grandest of symphonies. It awakens in one a new interest in his fellow-man and a sobering and devout sense of his relation to the universe.[9]

Although this article was over the head of the average reader, it was a sensation among such Tesla backers as Morgan, Kelvin, and those followers who hung on his every word and prophecy. It became the talk for both scientific and financial worlds and came in for acclaim from those more impractical, metaphysical groups, who, for a long time, had felt that Tesla was one of them.

Tesla went on to further dizzying predictions in still later articles in popular publications. He discussed interplanetary communications at some length, asserting positively that such communication was to become the dominating idea of the new century and that he had received a call from space which had originated on Mars, Venus, or some sister planet. There were a few noted scientists who were disposed to agree with him. He was quoted as saying that in the solar system there seemed to be only two planets capable of sustaining life such as ours, but that there could well be other forms of life on them, that perhaps there might be a form of existence which did not require nourishment such as we know it.

> Organic life might undergo . . . modifications, leading to forms which, according to our present ideas of life, are impossible. Changes could be gradual. . . . So I think it is quite possible that in a frozen planet, such as our moon is supposed to be, intelligent beings may still dwell in its interior, if not on its surface. . . . With the novel means proposed by myself . . . with an expenditure not exceeding two thousand horse-power, signals can be transmitted to a planet such as Mars with as much exactness and certitude as we now send messages by wire from New York to Philadelphia.[10]

All these dramatic predictions had stemmed from his Colorado experiments where he said he felt as if he were present at the birth of a new knowledge or the revelation of a great truth. He insisted that "there was present in them something mysterious, not to say supernatural."[11]

Little wonder that in the practical, scientific world the temple came crashing down about his ears. He was damned with words of warning that his recently published utterances had discredited him in the eyes of competent judges. Caustic criticism of the *Century* article appeared in leading scientific publications. Tesla was accused of seeking newspaper notoriety. His article reviewed the reasons for his trip to Colorado, saying it had been first to transmit power without wires, second to develop apparatus for submarine telegraphy, and third to study planetary communication. Scientific writers jibed: "On the first point, when we hear of it from the patent office, we will discuss it." On the second point they took the same stand: and as for the third, planetary communications, they warned that Tesla's writings must be read with extreme caution, that electrical experiments could only be judged by commercial success, and that Mr. Tesla's speculations were so reckless as to lose interest and his philosophy so ignorant as to be worthless.[12]

And because of his fantastic predictions, charlatans, crackpots, and the fanatics grabbed at Tesla's coat tails so that he became identified with them through no desire or intent of his own. This did little to enhance his reputation in scientific circles. He became a victim of guilt by association—an association of ideas rather than ideals. His remote personality, his Slavic features, with the common knowledge that he had his birth in the land of the Magyars, added color to the fabrications being built around him. He realized that this was not an asset in either financial or scientific circles and took every opportunity to deny his alliance with the occult.

At times he was unable to explain to those untrained in science the wonders of the future in such a way that they were able to share his excitement, but one writer declared that Tesla could make intelligible to the untrained the spiritual essence of his specialties and discoveries and went further to declare:

> To be with Tesla is to enter a domain of Freedom even more free than solitude because the horizon enlarges so. . . . With music he stood on terms of a musician; and one can imagine the kind of transcendental symphony that a mind of that calibre might compose.[13]

This writer stated that Tesla gave a total impression, of knowledge, not divided into chambers, but an "entire chrysolite of intellectual ability," and that when Tesla spoke one could see mankind rise as a Titan and look into the future when the human race was no longer forced to labor and one could see a day when

> rich and poor no longer meant difference of material conditions but of spiritual capacity and ambition—a time when intercommunication all over the earth should be immediate and universal and even when knowledge should be derived from sources now hardly imagined.[14]

All this poetic exuberance meant little to the financial magnates who preferred concrete results backed up with cold cash which would quite possibly arise from Tesla's frenzied efforts on behalf of his world broadcasting system. Morgan's $150,000 was woefully inadequate when Tesla considered all that must be done, but it was a start. He secured a tract of land on Long Island, about sixty miles from New York City, through an arrangement with James S. Warden. Tesla had pictured to Warden a glowing and convincing real estate boom in that site, employing several thousand people who would build their homes on the adjacent land. Warden cooperated to the extent of offering two hundred acres of land for the use of the scientist, twenty acres already cleared and with a well one-hundred feet deep. By July 23, 1901, work had started on the project with the roads cleared and the right of way in order.[15] Thus, within a little less than five months after the contract with Morgan was signed, work was started on Tesla's giant project.

Tesla's dreams included an industrial Utopian city. The project which began to take form was called Wardenclyffe. In drawing up the plans, Tesla enlisted the services of his friend Stanford White, who offered to underwrite the designing of the dream city and the tower. White already had behind him such monumental achievements as the Washington Arch in New York, New York University, the University of Virginia, and Trinity Church in Boston. The most important physical feature of the new project was the giant tower, which took shape despite hot discussions between the designer and his associates who knew how vulnerable such a structure might be to the winds, and Tesla who knew only what he must have to carry out his project. Again his weakness in architectural matters, which had been noted in Colorado Springs, loomed as an obstacle. During most of the two years of construction, Tesla commuted daily from the Waldorf to Wardenclyffe, accompanied by a Serbian manservant bearing a large hamper of food.

Among those working on the station was Bernard Arthur Behrend, one of the few who could speak Tesla's language of engineering. He had been in charge of an engineering firm which designed one of the motors for the Wardenclyffe plant and was recognized by Tesla for his ability and his keenness in grasping the requirements of the Wardenclyffe experiment. Later Behrend won the life-long respect and devotion of Tesla when, during litigation, he refused to testify for the company which employed him, because he was convinced they were defrauding Tesla in a patent suit.[16]

By the end of two years, a brick laboratory and a powerhouse were completed, dominated by a tower 187 feet high with a spherical terminal about sixty-eight feet in diameter. It was estimated that the equipment cost something in the neighborhood of $200,000.[17]

By spring of 1903 Tesla found himself in the awkward position of trying to vindicate his lack of results in view of the great expense. He tried repeatedly to explain to J. P. Morgan what the rising costs had done to his project. He wrote:

> You have raised great waves in the industrial world and some have struck my little boat. Prices have gone up in consequence, twice, perhaps three time higher than they were and then there were expensive delays, mostly a result of the activities you have excited.[18]

Tesla was certain, however, that if he only had more funds forthcoming his project could be completed by July. He had sold some personal property to raise additional funds, had borrowed $10,000 from the bank, and explained that financially he was in a dreadful fix.[19] By nature, Tesla was a proud man and it was distasteful and humiliating to beg, but he was desperate to save his dream and willing to bow low if by so doing he could save Wardenclyffe. At first, his pleas to Morgan were dignified and business-like, but when such appeals met only with a firm and often curt "no," his dedication to his messianic dream brought Tesla to his knees and to the point where he was willing to be spat upon and crucified for the sake of what he believed.

plained again and again that his world system was ...ed to make possible

> not only the instantaneous and precise wireless transmission of any kind of signals, messages or characters, to all parts of the world, but also the inter-connection of the system, telegraph, telephone, and other signal stations without any change in their present equipment.[20]

He further intended to prove that by its means a telephone subscriber here might talk to any other subscriber on the globe, by means of an inexpensive receiver no larger than a watch; that by this means one could listen anywhere on land or sea to a speech delivered or music played in any other place, regardless of distance. The earth, he declared, was a perfect natural conductor, available for all the innumerable purposes which human ingenuity had already found for line-wire.[21]

Nor were these the only benefits to be projected from this tower. Tesla had plans for secret government telegraph service, inter-operation of all the world's stock tickers, universal registration of time, global transmission of typed or hand-written letters, a universal marine service enabling navigators to steer without compass and to determine the exact location and speed of other vessels, to prevent collisions. He envisioned a system of world printing on land and sea, including the reproduction of photographs, drawings, and documents. This was only a start for his plans of world magnitude. He was prepared for the excitement he believed was coming when humanity would be "like an ant-heap stirred up with a stick."[22]

He maintained his constant insistance upon order and cleanliness at all times in his laboratory. This led to near disaster at the Wardenclyffe plant. An assistant had meticulously scrubbed the floor, but allowed some water to stand in the holes which had been drilled in the floor. When Tesla's financial secretary, George Scherff, who was helping in the laboratory, poured molten lead into the hole which retained some of the scrub water, an instantaneous explosion hurled bits of lead into his face. Tesla, standing nearby with a second ladleful, was also burned. Scherff was threatened with blindness, but eventually recovered. It seems remarkable that there were not more such incidents with the potentially dangerous experiments carried on in the Tesla laboratories. Only one other is recorded when a young assistant suffered X-ray burns, a medical hazard not fully understood at that time.

So engrossed was Tesla with the magnitude of his endeavors that he failed to safeguard his patent rights which were thus pirated on all sides. Economic factors which he could not fathom were gathering against him. Creditors clamored for money due them, but he had no time to be bothered and no way of obtaining more funds except through his increasingly persistent pleas to the adamant Morgan. He boasted that, "If you will imagine that I have found the stone of the philosophers, you will be not far from the truth."[23] Tesla regretted the fact that his enemies had represented him as a poet and a visionary, which put the necessity on him of immediately producing something of commercial value to refute their remarks.[24]

And finally the worries and frustrations grew more than he could bear. Tesla had written Morgan on the occasion of his New Year, stating that Canadian Niagara would furnish 10,000 horsepower for twenty years without charge if he would put up a plant there to transmit power without wires—and he again begged for financial assistance.[25] A blunt no was sent back immediately.

Tesla, in bitterness, replied:

> That was a nice letter to receive on my New Year! Had you at least waited till today, bad news travels fast enough. You wish me success. It is in your hands. How can you wish it?
>
> We start on a proposition everything duly calculated, it is financially frail. You engage in impossible operations, you make me pay double, make me wait ten months for machinery. On top of that you produce a panic. When, after putting all I could scrape together, I come to show you that I have done the best that could be done you fire me out like an office boy and roar so that you are heard

> six blocks away: not a cent. It is spread all over town, I am discredited, the laughing stock of my enemies.
>
> It is just fourteen months that the construction work on my plant was stopped. If I would have been helped at that time, three months more with a good force of men and now it would be paying ten thousand dollars a day. More than this, I would have secured contracts from governments for a number of similar plants. I am the discoverer of the principles and the inventor of all the essential devices and no one would have the slightest chance in competition with me. You have favored the schemers who have no knowledge or skill but merely the cursing sense of fraud to fool the world and to hurt my work by their incompetent attempts and far more than they ever could by success.
>
> . . . Men are like flies to you. . . . Please do not write to refuse. . . . I am pained enough as it is.
>
> Yours sorrowfully,[26]

But again, he was forced to beg, and again and again. He wrote:

> I am tired of speaking to pusillanimous people who become scared when I ask them to invest five thousand dollars and get the diarrhoea when I call for ten.[27]

And later:

> Have you ever read the book of Job? If you will put my mind in the place of his body, you will find my sufferings accurately described.[28]

Tesla explained that he could not develop the business slowly "in grocery-shop fashion." He wrote:

> . . . You told me at the outset that I should not ask for more but the work was of such transcending importance

> that I undertook to explain to you on your first return from abroad.
>
> . . . To achieve is one thing—To achieve at the right moment is another. That favorable moment is gone forever.
>
> . . . As soon as people find that messages, for, say, five cents a word can be transmitted to any distance, nothing will stop the demand for cheaper and quicker communication.

And he concluded, "Since a year, Mr. Morgan, there has hardly been a night when my pillow was not bathed in tears, but you must not think me a weak man for that. I am perfectly sure to finish my task, come what may."[29]

But again he was met with a firm refusal and he sent an angry reply:

> You are a man like Bismarck, great but uncontrollable. . . . But you are no Christian at all. You are a fanatic. Musoulman. Once you say, no, come what may, it is no.[30]

With a characteristic lack of business acumen, Tesla sometimes took the money which was given him to bolster his business ventures, where they could be established on a paying basis, and, instead, paid off past debts of honor. He was rapidly reaching the bottom of the barrel of his resources and soon became aware that he must seek a paying job.

In a desperate attempt to keep his head above water, he decided to yield to circumstances and to offer his services as a consulting electrician and engineer with his promise that any task which he accepted would be carried out thoroughly and conscientiously. To announce his availability, he issued, on January 1, 1904, an elaborate manifesto headed by the Latin phrase, *"Nihil in Sacculo quod non fuerit in Capite."* (There is nothing in the pocket that was not first in the head.)

With his typical disregard for diminishing resources, he issued it from the Waldorf, printed on the most expensive of vellum paper, and running to several pages. It bore a picture of Wardenclyffe Tower, listed ninety-three patents issued to Tesla, had a vignette of Niagara Falls, plus quotations from his lectures, and the envelope was personally sealed with red wax.[31]

Scherff had long advised him to lay aside his work temporarily for more lucrative channels, but up to this time, Tesla had refused. These were picayune matters which must not obstruct his plans—plans which were eventually sure to produce millions, given their time. Other friends continued to urge him to do the practical thing, to concentrate on any one of his ideas which might make him a millionaire with funds to complete his dreams. But Tesla, the impractical, in spite of his manifesto, went on dreaming—and the millions failed to appear.

He approached the great Morgan with still another plea and, failing that, requested that he be released of all obligations to Morgan, that his assignments of patents be given back to him. He begged Morgan to consider what he had already invested as a generous contribution leaving it to the integrity and ability of the inventor to work out the best results for both of them—and proposed a lecture course to make more money.[32]

The Morgan reply was not long in coming and stated in no uncertain terms that they had a contract, that he, Morgan, had fulfilled his part and did not consider it unreasonable to expect Tesla to carry out his.[33]

Tesla made a hasty and acrimonious reply:

> Owing to a habit contracted long ago, in defiance of superstition, I prefer to make important communications on the 13th of each month, but my house is afire and I have not an hour to waste.
>
> I knew that you would refuse. What chance have I to land the biggest Wall Street monster with the soul's "spider thread?"

> You said that you had fulfilled your contract with me. *You have not.*
>
> . . . You are a big man but your work is wrought in passing form. Mine is immortal.
>
> . . . Six thousand million dollars is invested in enterprises based on my discoveries in electrical science today.
>
> . . . When we entered our contract I furnished:
>
> 1. patent-rights,
> 2. my ability as an engineer and electrician
> 3. my good will.
>
> You were to furnish:
>
> 1. money
> 2. your business ability
> 3. your good will.
>
> I assigned patent rights which, in the worst case, are worth ten times your cash investment. You advanced the money, true, but even this first clause in our contract was violated. There was a delay of two months in furnishing the last $50,000, a delay which was fatal. . . . You discredited me.[34]

But Morgan sailed for Europe without granting Tesla a requested interview.

Two months later Tesla approached Morgan with another proposal—Morgan to exchange his patent interest for one-third of the stocks and bonds in companies which Tesla planned to organize.[35] Morgan countered that he was quite prepared to take one-third of any securities in any company in which Tesla might have an interest or which might be under his control as full settlement for the loan—*provided* that there be secured other capital adequate for the purpose of carrying out the business.[36] This proposal was obviously untenable.

Tesla realized that a project abandoned by Mr. Morgan was "dead as a door nail," and while he kept the gentleman's agreement that all their transactions be kept confidential as between friends, it appears that Morgan's refusals were bruited abroad so that Tesla was unable to interest other possible investors. Capitalists were unwilling to rush in where a Morgan feared to tread.

As far away as Colorado, the creditors were clamoring. Tesla had been threatened with a suit by the Colorado Springs Electric Company, and, having failed to answer the summons, his property was advertised for sale. Through one of his small windfalls, he gathered enough money to cancel the bill and the station again became his property.

Out on the Knob Hill prairie the wind tore at the tar paper and curious children walked by the deserted laboratory that took on the mystery of a haunted house. Pearl Overton, a little girl living on nearby Custer Street, wheeled her little sisters, Ruth and Nellie, in their baby buggy out on the commons to pick flowers. Sometimes they stopped and asked the laboratory caretaker if they might look in. The walls were a maze of wires, incomprehensible to little girls with bouquets of bluebells held tightly in their hands.[37]

In 1904, presumably on Tesla's order, his Colorado Springs property was sold to Chalmers E. Maddocks, who was building a house in Ivywild and could use the materials. The piping went to St. John Plumbing Company and the apparatus was stored. Some say it was placed in the Colorado College powerhouse and was there for years, but it is possible this might have been confused with the equipment belonging to Lee De Forest, radio experimenter.[38]

Newspapers reported that the laboratory caretaker, Carl J. Duffner, had filed suit against Tesla for back wages and that the property was sold to satisfy the judgment. Curiously, it is also recorded in newspapers that Duffner was a bidder for the equipment and that he purchased an electrical transformer worth hundreds of dollars for a bid of ten. The sheriff's hammer had hardly fallen when the transformer was attached by an officer acting under instruction from Charles A. Pollen, who claimed

$80.00 in taxes, thus throwing the tax on the purchaser—but he still had a bargain.[39]

Even more strangely, the judgments, suits, and tax transactions which harried Tesla and were reported in Colorado Springs newspapers are nowhere to be found in records of the county or district court. It was said that Tesla returned to Colorado Springs in September, 1905, to answer the court action, accompanied by his attorney, M. L. Brown, but it is not certain if this is true.[40] It appears unlikely from the heading of letters written in New York during that period.

It is definite that yapping creditors dogged his heels vociferously and that he had nothing with which to placate them. His Wardenclyffe laboratory had come to an abrupt halt.

Upon the death of the senior Morgan, Tesla lost no time in approaching the son and was able to borrow $15,000 for nine months, at six percent interest[41] and in two months borrowed another $5,000 at the same interest rate.[42] For a time, somehow, Tesla met his payments on the interest, but there came letters begging for an extension of the loan, which was granted.

On one occasion Tesla wrote of his turbines which were all ready to be shipped to Germany, but, while they stood at the Edison plant ready for shipment, he lacked the $5,000 required. Young J. P. Morgan, Jr. advanced the necessary loan, to be paid within six weeks at the customary interest.[43] Actually, the turbines, when tested, proved to be full of holes and had to be reconditioned. When Tesla wrote again, asking for further funds, young Morgan refused. Tesla's dream had become a nightmare.

And so, the Wardenclyffe Tower began to take a ghost-like appearance. Workmen who had been silent about what they were doing simply disappeared. Almost overnight the machinery grew quiet. A night watchman patrolled the grounds, then he, too, left the Oz-like structure standing alone and forgotten by the countryside.

A few still remembered and predicted that the dream might yet become a reality and the site might become a mecca for admiring scientists. But weeds grew over the Wardenclyffe subdivision, curious boys climbed into the tower until it grew rotten, and parents declared it an eyesore and a public danger. The old tower was the uninscribed and forgotten monument to a heartbreak dream.

Tesla wrote to his close associate, Scherff, after going by his abandoned tower years later, "I did not exactly cry when I saw my place after so long an interval, but I came very close."[44]

Notes

1. Microfilm letter, Tesla to Westinghouse, January 22, 1900, Library of Congress.

2. Ibid., Tesla to J. P. Morgan, December 10, 1900.

3. Ibid., Tesla to C. W. Steele (Morgan's secretary), March 1, 1901.

4. Ibid., Tesla to J. P. Morgan, September 13, 1901.

5. Ibid., R. U. Johnson to Tesla, Undated.

6. Nikola Tesla, "The Problem of Increasing Human Energy," *Century Magazine*, June 1900, 175-211.

7. Ibid.

8. Ibid.

9. Ibid., 312.

10. Nikola Tesla, "Talking with Planets," *Colliers Weekly*, vol. xxvi, no. 19, February 9, 1901, 4-5.

11. Ibid.

12. "Task's Science," *Popular Science*, February 1901, 436-437.

13. J. Hawthorne, "Personality of Tesla," *Current Literature*, November 1900, 222.

14. Ibid.

15. Microfilm letter, Scherff to Tesla, July 13, 1901, Library of Congress.

16. Interview with Kenneth M. Swezey.

17. *Literary Digest*, September 8, 1917, 25.

18. Microfilm letter, Tesla to J. P. Morgan, April 8, 1903, Library of Congress.

19. Ibid.

20. Nikola Tesla, "My Inventions," in *Electrical Experimenter*, July, 1919, 177.

21. Ibid.

22. *Nikola Tesla, 1856 – 1943, Lectures, Patents and Articles*, (Beograd, Yugoslavia: Nikola Tesla Museum, 1956), A-161.

23. Microfilm letters, Tesla to J. P. Morgan, October 15, 1903, Library of Congress.

24. Ibid., December 11, 1903.
25. Ibid., January 13, 1904.
26. Ibid., January 14, 1904.
27. Ibid., April 1, 1904.
28. Ibid., April 2, 1904.
29. Ibid., October 13, 1904.
30. Ibid., October 17, 1904.
31. "A Striking Tesla Manifesto," *Electrical World and Engineer*, vol. xiii, no. 6, February 6, 1904, 256.
32. Microfilm letters, Tesla to J. P. Morgan, December 16, 1904, Library of Congress.
33. Ibid., J. P. Morgan to Tesla, December 18, 1904.
34. Ibid., Tesla to J. P. Morgan, December 19, 1904.
35. Ibid., February 15, 1906.
36. Ibid., J. P. Morgan to Tesla, February 16, 1906.
37. Interview with Pearl La Fair.
38. Interview with Arthur Baylis.
39. *Colorado Springs Evening Telegraph*, March 22, 1906.
40. *Colorado Springs Gazette*, September 6, 1905.
41. Microfilm letters, Tesla to J. P. Morgan, May 20, 1903, Library of Congress.
42. Ibid., July 23, 1913.
43. Ibid., J. P. Morgan to Tesla, January 5, 1914.
44. Ibid., Tesla to Scherff, July 13, 1913.

Nikola Tesla with a wirelessly-illuminated glass bulb, ca. 1919. *Sarony studio photograph, from Kenneth M. Swezey Papers, Archives Center, National Museum of American History, Smithsonian Institution (AC0047-0000014).*

Development of Turbines

The necessary closing of the Wardenclyffe project by signaled the death of ideas in the mind of Tesla. There were moments when he realized the bitter truth that he was years ahead of his time and might have to wait for time to catch up. There were other periods of optimism when he refused to admit defeat. The problem now was the interlude until he could take up the incompleted project again.

He concerned himself with further developments in his alternating current polyphase system. His mind turned to rotary engines and to steam turbines of an improved type.

One of Tesla's original turbines, in 1906, was a six-inch, ten-pound machine employing a thirty horsepower motor. To signal its success, Tesla ordered new stationery with letterheads and envelopes bearing the slogan, "Twenty horsepower per pound."[1] His largest turbine he considered a triumph in engineering skill. This turbo-dynamo, the largest ever made, developed 30,000 horsepower and was made from his plans worked out on paper. It had never been tested until it was erected at Niagara Falls and worked perfectly from the first turning on of steam.[2]

By contrast, in 1912, he developed the smallest model turbine. There was still a member of the Czito family on the Tesla staff, Julius, son of the Kolman Czito who had thrown the switch in Colorado Springs. It was Julius who assisted him with the small turbine which was the talk of engineering circles.

Of this small turbine, a "powerhouse in a hat," Tesla said:

> What I have done is to discard entirely the idea that there must be a solid wall in front of the steam and to apply in a practical way, for the first time, two properties which every physicist knows to be common to all fluids but which have not been utilized. These are adhesion and viscosity.

> To say nothing of it being a new application of mechanical principles, it has many decided advantages. First of these is its simplicity. It is comparatively inexpensive to construct, because nothing but the bearings need be accurately fitted, and exact clearances are not essential. Then there is nothing to get out of order and the disks can easily be replaced by any competent mechanic. It can be reversed without complex or cumbersome apparatus—all that is needed is a two-way valve to let the steam in at one side or the other, as desired. Reversing an ordinary turbine is next to impossible.[3]

Daytime was filled with work, but the dinner hour usually found Tesla in the Rose Room of the Waldorf, at his accustomed table near the wall, a table large enough for two, but always set for one. His youthful appearance denied his years. His face had a timeless quality. His black hair bore no evidence of stress or time. Only his eyes, unaccountably sad as he faced the empty chair opposite him, gave evidence of the self-imposed loneliness of the celibate existence of one who lived on an inaccessible plane.

He could have been a broken old man from the weight of disappointments, yet the future held so much promise for him, his life was so full that he was unconscious of the passage of time and unaware that he should be growing old. He told a reporter of one Tesla grandfather who had lived to be over 118 and of another who won a foot race at seventy-three. He seemed aware that in the intensity of his interests lay the secret of his vitality.

> I went into electrical science years ago because I thought in that direction I was to solve the problem I have been working on all my life—the production of an engine sufficiently light and powerful to operate the ideal flying machine. All my work in the wireless transmission of power which has attracted more public attention than anything else I have ever done was toward that end. I expect to be young enough at 104 to make a flight in it.[4]

To another writer, Tesla revealed that he had a new turbine which could drive the *Lusitania* across the Atlantic at a speed of fifty miles per hour, using 90% of the turbine's energy. He described the operation as one which transmitted steam around the spiral circuits and imparted energy all along its track, while in the old type, steam made only one circuit and, when it escaped, still contained wasted energy. He said he could guarantee to put his innovation into a Wright biplane, but added that no one could sit in the plane at such speed. Tesla further stated that he expected to develop a new type of flying machine that could be driven by means of his turbine at almost the speed of a bullet.[5]

During this period he tried to interest various manufacturers in the turbine. He went to work as a consultant for Allis Chalmers, but again the old problem of working with people posed an impossibility. They complained that his designs were only in his head, and they couldn't work from theory. Tesla, on the other hand, felt they should be able to do this if they were qualified engineers. He gave no margin for human frailty or lack of engineering imagination. In turn, they were unable to discount the behavior of genius. The result was the usual stalemate. Tesla threw up his hands and left before important tests were completed. He was growing more rigid and increasingly unable to compromise. The practical world crowded him out.

While he occupied himself with the conserving of energy and power of steam to do the work of the world without waste, the world prepared for its greatest waste. No one felt the omen of the clouds of war gathering over Europe in 1914 more than did Tesla. He sensed that it would be his own countrymen, the Serbs, who would be the first pawns in the devastating game and commented: "Civilization alone is evidently insufficient for insuring peace on earth." He predicted that this war which had not yet touched America would be a vast terrestrial upheaval, "as if gigantic forces were unchained, threatening the entire globe."

Typically, he simplified the situation with a mathematical equation, stating that war is essentially the manifestation of energy involving the acceleration and retardation of a mass by force. He foresaw the fall of Austria and the probable loss of

Alsace-Lorraine. He spoke of the hypnotic influence of military training, eliminating individual initiative. This, he said, was the formidable engine which Germany had perfected. He spoke of the blunder of the violation of Belgian neutrality and of the blind faith of the German offensive. He predicted that war could only be ended by exhaustion. The very intelligence and resourcefulness of the Germans was a threat—"they were fully capable of making two blades of grass grow where one had grown before." The responsibility for the calamity, Tesla placed at the door of modern science, but he was convinced that science could also undo the Frankenstein monster which it had created.[6]

He pondered on the question of whether the world's future development would be in the direction of the artistic and the beautiful or the scientific and the useful, and concluded that art must be sacrificed to science, therefore the rational Germans represented the nearest approach to the humanity of the future, with the grim forecast that "a world of bees will be the ultimate result."[7]

It was impossible for most Americans to follow Tesla's thinking. Already they were beginning to form the idea that this was a war to end all wars, theories or no theories. They were unaware that his analogy of humming bees was already a reality and not merely a metaphor. Already the hum of world-wide unrest was beginning to swarm toward the hive of communism.

Notes

1. John J. O'Neill, *Prodigal Genius*, (New York, N.Y.: Ives Washburn, Inc., 1944), 218.

2. Frank Parker Stockbridge, "The Tesla Turbine," *The World's Work*, March 1912, 543-548.

3. Ibid.

4. Ibid.

5. A. L. Benson, "Nikola Tesla, Dreamer," *The World Today*, February 1912, 1763-1767.

6. *Nikola Tesla, 1856 – 1943, Lectures, Patents and Articles*, (Beograd, Yugoslavia: Nikola Tesla Museum, 1956), A-162-171.

7. Ibid.

Nikola Tesla's Colorado Springs laboratory from the rear with the Pikes Peak range in the distance, late December 1899. This may be one of the photographs of the lab that Tesla described as being made by moonlight, utilizing a long time exposure. Tesla wrote in his notes, "The above photographs satisfied fully the novel pleasure of taking pictures by the fascinating moonlight of Colorado and attention was then turned to less agreeable but more useful work." The writing on the photograph reads, "Experimental station fully developed. Activity (power delivery) one hundred thousand horsepower." *Courtesy Nikola Tesla Museum, Belgrade, Serbia, (006-04-12).*

Edison and Tesla To Get Nobel Prizes

TESLA'S DISCOVERY NOBEL PRIZE WINNER

Transmission of Electical Energy Without Wires, Which Affects Present-Day Problems

TO ILLUMINATE THE OCEAN

Scientist Says Collisions Will Be Avoided and Unlimited Water Drawn to Irrigate Deserts.

New York Times, November 6 and 7, 1915.

The Mystery of the Nobel Prize

Disappointment continued to dog Tesla in 1915, when a long-deserved honor miscarried. The full story will probably never be known and every researcher seems to have his personal opinion of what happened.

Through a dispatch from Reuter's, by way of *The London Times,* news came to New York that Tesla and Edison were to be chosen that year to share the Nobel Prize in physics. *The New York Times* made a front page insert on November 6th, repeating the announcement.[1] On the following day, the paper printed an interview with Tesla on the subject. Tesla declared that he had not received notification of such an honor, but said that such an award might be for a recent discovery in the transmission of electrical energy without wires. "This discovery," said Tesla, "means that electrical effects of uniform intensity and power can be produced so that not only can energy be transmitted for all practical purposes to any terrestrial distance but even effects of cosmic magnitude may be created."[2]

Tesla grew enthusiastic in his prediction to the reporter who quoted:

> We can illuminate the sky and deprive the ocean of its terrors! We can draw unlimited quantities of water from the ocean for irrigation! We can fertilize the soil and draw energy from the sun!

Tesla further predicted that the day would come when all battles would be waged with electrical waves instead of explosives.

When asked to conjecture on what discovery of Edison would be honored, Tesla did not comment except to say that Edison was worthy of a dozen Nobel Prizes.

In Omaha, Thomas A. Edison, returning from a journey to the Pacific coast to the Panama-Pacific Exposition, was reached by the press for his reaction to the news. He appeared surprised when shown the London dispatch. He also declared that he had received no official notice. He declined to comment further.

Tesla must have been jubilant over his recognition, for he wrote immediately to his friends, the Robert Underwood Johnsons, who had been among the first to congratulate him. He thanked them for their joy in his good fortune. Then he added a sobering note, separating the value of the prize from the value of the achievement, saying:

> In a thousand years, there will be many recipients of the Nobel prize. But I have not less than four dozen of my creations *identified with my name in technical literature.* These are honors real and permanent, which are bestowed, not by a few who are apt to err, but by the whole world which seldom makes a mistake, and for any of these I would give all the Nobel prizes during the next thousand years.[3]

Then his inherent optimism of the fortune that lay always around the corner came to the fore and he concluded the letter in a jovial mood:

> Josie will never have the chance of turning me away as a beggar, but I shall give her soon, the opportunity of slamming your door in the face of a millionaire.[4]

The Electrical World of New York, on November 13th, spoke of the satisfaction to the engineering world to have the honor given to these two outstanding men. Both Tesla and Edison were lauded for their accomplishments.

There were many who remembered a Tesla patent announcement on the 13th of the previous month and were convinced that that particular patent had featured in the impending award. Tesla had been quoted in *The Times* as saying:

> It is possible to project the human voice not only for a distance of 5,000 miles but clear across the globe. I demonstrated this in Colorado in 1899.[5]

He grieved that the world did not yet ur principles of his transmission even after fifteer same October article he had explained how the sy work:

> The plant would simply be connected with the telephone exchange of New York City and a subscriber will be able to talk to any other telephone subscriber in the world, and all this without any change in his apparatus.[6]

The article further quoted his predictions to transmit pictures, and project images, and stated that he could facilitate the work of the press so that a picture from the battlefield of Europe could reach New York in five minutes.[7] He explained:

> The current passes through the earth, starting from the transmission station with infinite speed from that region and, slowing down to the speed of light at a distance of 6,000 miles, then increasing in speed from that region and reaching the receiving station with infinite velocity.
>
> It's a wonderful thing. Wireless is coming to mankind in its full meaning like a hurricane, some of these days. Some day there will be, say, six great wireless telephone stations in the world system connecting all the inhabitants of this earth to one another not only by voice but by sight.[8]

Remembering these predictions by Tesla, many of his friends were convinced that the award was certain and deserved.

The events of the following weeks were confusing to the two inventors who were now centered in the public spotlight, and even more confusing to the scientific world. The mystery of what actually happened has been a stumbling block for researchers. One of the most reliable of biographers reported the incident as occurring in 1912, but this obviously could not be correct since all press releases point to 1915 as the date. Numerous others have accepted the 1912 date blindly. The year 1915 was a time of war and there was also some question whether or not the Nobel Prizes would be awarded in many cases.

However, on November 13, 1915, *The Electrical World* also announced the American inventors as chosen candidates for the honor. On November 14th, consternation reigned for those who happened to be following the story. According to a Reuter dispatch from Stockholm, the Nobel prize for physics was to be divided between Professor William Henry Bragg of the University of Leeds and his son, William Lawrence Bragg, of Cambridge University for research in the structure of crystals by use of Roentgen rays.[9]

On November 15th, *The London Times* repeated the dispatch concerning the Braggs.

It is possible that the sudden switch of awards went unnoticed by the casual readers and others were so sure of the outcome that they failed to comprehend what had happened. Why the reversal of decision had taken place, no one knew. A biographer of Tesla, who had been a close friend, wrote many years later that Tesla had refused the honor, stating that he could not bring himself to share the prize with Edison who was an inventor, while he, Tesla, considered himself a discoverer. The story has been repeated and repeated, but it is entirely out of character with Tesla's behavior pattern. He was a gentleman under all circumstances, and the enthusiasm with which he paid tribute to the genius of Edison, when the ill-starred first announcement had come, would scarcely warrant such a reversal of attitude. Rather, it would have been more in the pattern of Edison to have refused. There had been strained relations between the two since the day Tesla had tipped his hat and left the Edison laboratory to become his bitter opponent with Westinghouse. Edison might yet have smarted under the necessity of having to pay for license to use the Westinghouse patents. To have deprived Tesla of $20,000 in his hour of need might have been a factor if Edison had chosen to refuse, and it was quite in keeping with his sardonic and sadistic brand of humor. But it is quite possible that neither one of these conjectures carries the full story, or is even partly true.

According to correspondence with the Royal Swedish Academy of Science, the statutes of the Nobel Prize Foundation preclude any information concerning Nobel Prizes or any proposals which may or may not have been made to the Academy. However, they offer this ray of explanation. They say:

> Any rumor that a person has not been given a Nobel Prize because he has made known his intention to refuse the award is ridiculous. A decision by a prize awarding body to create a laureate, so to speak, does not depend on the opinion he may have on the act. Should a person have the high distinction of being created a laureate in this way, only to refuse to accept prize money, the medal and the diploma, this will, of course, be officially recorded, but it does not make a particle difference to the fact that he has actually been made a Nobel Laureate by the prize awarding body in question.[10]

Although such a statement is somewhat noncommittal as to what really happened, it may merely indicate nothing happened, or it may be a smoke screen. There are some who believe that a news leak of the intended winners caused the committee to change the original plans.

It seems strange that although Bragg was proclaimed the winner by November 15th, newspaper writers were among those who assumed Tesla had been honored and *The New York Times* printed an article on December 8th, about a new wireless device. The writer, in the opening sentence, referred to Tesla as the winner of the 1915 Nobel Prize for physics. Evidently he did not read his own paper of three weeks previous or else did not believe it.[11]

Ten days later on December 18th, *The Literary Digest* announced that Tesla and Edison and Professor Theodore W. Richards of Harvard would be the three Americans honored by Nobel Prizes. Either the article had been written too far in advance, or the writer was still unaware that the prize had been given over a month before. The article was lengthy and lauded all three Americans.[12]

How deep the cut was for Tesla will never be known, but it must have been painful. That he either could not—or at least did not—discuss his disappointment with his closest friends, the Johnsons, is indicated from a letter which he received in March of 1916, five months after the award to the Braggs. Johnson had written facetiously:

> When that Nobel prize comes, remember that I am holding on by the skin of my teeth, and desperately in need of cash.[13]

Once on a Serbian hill, Tesla had watched a snowball pick up speed and dimension as it careened dangerously down a slope, revealing the inevitable destruction of accumulating force. There was a strange analogy in this remembered incident to all the misfortunes that were snowballing upon him. Each rebuff added momentum to the avalanche that seemed to be descending upon him.

Notes

1. *New York Times*, November 6, 1915.
2. Ibid., November 7, 1915.
3. Microfilm letter, Tesla to Johnson, November 10, 1915, Library of Congress.
4. Ibid.
5. *New York Times*, October 3, 1915.
6. Ibid.
7. Ibid.
8. *New York Times*, November 14, 1915.
9. *London Times*, November 15, 1915.
10. Letter from Dr. Erik Rudberg, Stockholm 50, Sweden, Secretary of Royal Academy of Science.
11. *New York Times*, December 8, 1915.
12. *Literary Digest*, December 18, 1915, 1426.
13. Microfilm letter, Johnson to Tesla, March 1, 1916, Library of Congress.

Nikola Tesla was never awarded the Nobel Prize, however the significance of his scientific contributions have been commemorated with international currency and stamps. Some examples are pictured on the facing page.

НАРОДНА БАНКА
СРБИЈЕ
100
T = Wb/m²
сто динара
sto dinara
NARODNA BANKA
SRBIJE
100

NIKOLA TESLA
30

50
НИКОЛА ТЕСЛА
1856-1943
JUGOSLAVIJA

NIKOLA TESLA
ЈУГОСЛАВИЈА
75
ПАРА

250

GHANA
C1000

INDUCTION MOTOR
USA
Nikola Tesla
20c

NIKOLA TESLA
GHANA
C1000

Thomas Alva Edison, namesake of the Edison Medal. The Edison Medal has been awarded by the American Institute of Electrical Engineers (now the Institute of Electrical and Electronics Engineers—IEEE) since 1909. Nikola Tesla was presented the medal in 1916, "For meritorious achievement in his early original work in polyphase and high-frequency electrical currents." *Detroit Publishing Co. photograph, Library of Congress Prints and Photographs Division Washington, D.C., (LC-DIG-det-4a25922).*

The Award of the Edison Meda

To most of those who saw the lonely figure of Tesla on the New York streets during this period of rejection, he was a solitary wanderer, feeding the hungry pigeons in front of St. Patrick's Cathedral, the Public Library, or in Bryant Park. Tesla sought out the wounded birds and took them home with him, perhaps identifying himself with the injured ones with their broken wings—with all those who are meant to soar but are earthbound. In his own need to be needed by some living thing, Tesla made his daily rounds to his "feeding stations," and even enlisted the help of friends to take care of the helpless birds.

One such friend was Tesla's close associate, Bernard Behrend, who felt the sting of the neglect of Tesla and realized with bitterness that the scientist was fast becoming a man forgotten. Although Behrend, himself, was a likely candidate for the Edison Medal to be given in 1917 by the American Institute of Electrical Engineers, he persuaded the organization to confer this long overdue honor on Tesla.

According to some, the suggestion that he be awarded the medal was met by Tesla with some reluctance, either for the reason that the honor was so late in coming or for the reason that it bore the Edison name.[1]

Whatever his true feeling might have been, Tesla finally agreed to accept the medal. The meeting was held in the auditorium of the Engineering Society building in New York and the distinguished guests filed in to pay their honors to the recipient of the Edison Medal. One biographer maintains that just before Tesla was to speak, it was learned that he had disappeared and that he was found by Behrend, feeding the pigeons at the Public Library. Behrend had known quite well where to look for his friend. He led him gently back to the honors dinner.[2]

One of Tesla's most entrusted confidants, Kenneth Swezey, avers that this legend is not entirely accurate, but that Tesla had decided to leave the gathering and was caught by his coat tails by Behrend, just as Tesla was hailing a taxi.[3] The dinner went on, a bit behind schedule, with Behrend doing the honors of presenting the speaker of the evening. In making the presentation Behrend said:

> We beg you to cherish this medal as a symbol of our gratitude for a new creative thought, a powerful impetus, akin to revolution, which you have given to our art and to our science. You have lived to see the work of your genius established. What shall a man desire more than this? There rings out to us a paraphrase of Pope's lines on Newton: "Nature and Nature's laws lay hid in night." God said, "Let Tesla be, and all was light."[4]

Tesla had not prepared the brief response which was expected of him. Instead, he rambled in a discussion of electrical science and his early memories of it, like a garrulous old man, unmindful of his wearying audience. It has been said that he held the honorary medal of that evening in such little esteem that at one time, unable to pay his stenographers, Tesla offered it in payment for secretarial services.[5] If so, the offer must not have been accepted, for the Edison Medal, along with other awards, was found by Kenneth Swezey and others who later went through the possessions in Tesla's estate.

Tesla mentioned in an autobiography that his citizenship papers were confined in his safe, but that his medals, honors, diplomas, honorary degrees, and other distinctions were packed away in an old trunk.[6] Perhaps this was true at one time, but as he grew older and the honors grew fewer, he must have come to place them in higher regard and so transferred them to the safe alongside his citizenship papers, where they were found by Kenneth Swezey and those who sorted the material things that comprised Tesla's meager estate.[7]

Although he inwardly appreciated the sporadic honors, which still came, he was unable to balance them with the pyramiding disappointments. Tesla said little more of his being robbed of the Nobel Prize, but it was a wound he would carry always.

Not the least of his heartaches came with the total demolition of his Wardenclyffe Tower. Although it had long since been abandoned, it was still a part of a dream he had not relinquished.

War, with its breeding of suspicion, fears, and hysteria, pointed to the tower as a possible point from which German spies might be operating. Nearby residents wondered if the rumors of renewal of experimental work in the old tower might not, after all, be valid and if the work was being carried on by enemy spies. Tesla's acquired citizenship did not preclude the stigma of his foreign birth. Foreigners were foreigners.

It was reported that on September 1, 1917, German spies had been spotted sending information of troop movements from the tower.[8]

It was only a week later that announcement was made of the tower's total destruction by the United States government.[9] Many had predicted that the weathered old structure, built from the plans of Stanford White, would scatter like a handful of matches from the ordered blast of dynamite. It was a welcome demolition to the neighbors, and a death sentence they felt long overdue. To their surprise, however, instead of flying into bits, as everyone had anticipated, the old tower "merely lay over on its side, like a dying monarch, dignified to the last."[10]

Tesla resented the implications of his disloyalty to his adopted country which was implied from the destruction of his property. He always denied that the government was actually responsible for the demolition. If such had been the case, he insisted, then he would have been reimbursed with a large sum of money as recompense for the great fortune which he had put into it. He further declared that it was in the interest of the United States government to preserve the structure as a means of locating enemy submarines. Tesla felt that there was a personal motive involved. At the time of the tower's destruction, he made no public protest, feeling it was improper to dwell on personal tribulations when the whole world was so troubled, but two years later he wrote:

> I am unwilling to accord some small minded and jealous individuals the satisfaction of having thwarted

> my efforts. These men are to me nothing more than microbes of a nasty disease. My project was retarded by the laws of nature. The world was not prepared for it. It was too far ahead of time but the same laws will prevail in the end and make it a triumphal success.[11]

The suspicions fed upon themselves and the honors grew fewer and fewer, but some old friends did not forget. Robert Underwood Johnson was one of those who never came to disregard the importance of Tesla, as a friend, as a scientist, or as a patriotic citizen. Johnson urged Tesla to sit for a bust by the great Yugoslavian sculptor, Ivan Mestrovic. Johnson wrote:

> I wish you would sit for your portrait bust. My dear Nikola, you are not going to live forever and when you go you will leave behind you nothing but a few photographs to tell the world how you look. . . . You are sure you will no longer exist, but I, being sure, would like to be able to say, "That was my friend, Tesla." And maybe after all it is because we inspire to go on and do likewise that we put up effigies. Mestrovic is at 42 Washington Mews, near 8th St. Pull yourself together and give him and yourself a chance.[12]

Mestrovic was a famous sculptor, even at that date, and he was the one logical person to do the portrait bust of Tesla. The two had so much in common. However, there appears no record of Tesla's consent, Mestrovic was a Yugoslav patriot who held ideals and traditions of his country dear. Both men felt the worth of the individual. Mestrovic had once said, ". . . in no age has so much been talked and written about individuality as in the present one and in no age have these qualities been so visibly lacking."[13] He also declared: "Happy are those who live forever. They had a reason for being born."[14]

The two men were similar in their approach to their chosen fields. Like Tesla, the sculptor had a vision even before he set his chisel in stone, that within it was the figure he wanted to make, not only in rough outline but precisely and in detail from all sides. To both, once the vision was conceived, it was already completed in its total harmony, "without the hand or the eye taking part." Little wonder that Mestrovic seemed to Johnson the ideal sculptor to capture the essence of Tesla in stone.

Although the honors grew fewer, Tesla continued to issue predictions which startled laymen and scientists alike, only more and more they were only forecasts, without solid evidence of machinery to back them up. He had long advocated the uses of electricity for a dry bath of "cold fire," using millions of volts of power to throw off the dead skin, scale and dirt, a treatment which was considered highly invigorating and some predicted that every home would soon be equipped with a huge Tesla coil for such a purpose.[15]

These were days when Tesla sat quietly in his study reviewing the changing world and foreseeing still more changes to come. He wore his somber clothes and in the midst of luxury, lived meagerly, abstaining from all beverages save milk and water. He never dared remove his gaze from the incredible drama of life. He still felt that the greatest ills from which humanity suffered were due to the inability of individuals and nations to come to close communication. He was increasingly convinced that:

> When wireless is perfectly applied and the whole earth will be converted into a huge brain, which in fact it is, as all things being particles of a real and rhythmic whole, we shall be able to communicate with one another instantly, irrespective of distance. Not only this, but through television and telephone we shall see and hear one another as perfectly as though we were face to face . . . and the instruments through which we shall be able to do this will be amazingly simple compared with our present telephone. A man will be able to carry one in his vest pocket. . . . Perhaps the most valuable application of wireless energy will be the propulsion of flying machines, which will carry no fuel and will be free from any limitations of the present airplanes and dirigibles.

> . . . It is more than probable that the household's daily newspaper will be printed "wirelessly" in the home during the night.
>
> The problem of parking automobiles and furnishing separate roads for commercial and pleasure traffic will be solved. Belted parking towers will arise in our cities, and the roads will be multiplied through sheer necessity, or finally rendered unnecessary when civilization exchanges wheels for wings.[16]

He mentioned the potentialities of solar heat as a practical source of power for heating homes and factories.

Tesla considered one of the most profound portents of the future to be the inevitable emergence of women. He was convinced that a new sex order would arise with the female as the superior, that female capacity would be expanded, and that women would not only be well educated, but even better educated than men, since their faculties had so long lain dormant and therefore had become more intense and powerful.

Tesla felt that such a development was not without peril, however. He feared that with the acquisition of new fields of endeavor, women would finally dissipate their feminine sensibilities and choke out completely the maternal instinct so that marriage and motherhood might become abhorrent to them and human civilization would draw closer to that of the bee.[17]

Except for an occasional startling prediction, Tesla became more and more of a recluse. He scarcely mentioned any more the tragedy of his Wardenclyffe dream. Often he was subjected to ridicule by the press but he went on to predict houses lighted and powered by wireless, as well as such power applied to means of transportation. He had already designed an automobile to be propelled on water, or through air, by jets. The British Navy was experimenting with a warship without a crew, controlled by wireless, similar to the model Tesla had demonstrated long before in Madison Square Garden. He predicted flivver airplanes which would buzz directly upward or dart across the country. He invented an auto speedometer which was put into immediate use, and he made a pocket-sized revolution counter.

Although he could foresee jet-driven airships to cross the ocean, he drew the line at moon rockets.[18] He insisted:

> No rocket will reach the moon . . . save by a miraculous discovery of an explosive far more energetic than any known. . . . it would still have to be shown that the rocket machine would operate at 459 degrees below zero, the temperature of interplanetary space. . . . [it would] mean starting at an initial velocity of 6.8815 miles per second. Even if the rocket weighed a pound for every pound carried, 52693 b.t.u. must be developed for every pound of fuel.[19]

Tesla felt that a motor driven by atomic power would be irrational, for it would take far more energy to break up the atomic structure than could be rewarded by doing more useful work. Indeed, he considered atomic energy more impossible than the practical application of a perpetual motion machine. He did not believe in the existence of electrons and was convinced that radium was not a generator but a transformer of energy, coming from cosmic rays.

Still closest to his heart was his theory of the principle of the rotating magnetic field, and he would not give it for a thousand inventions "designed merely as mechanical contraptions to deceive the eye and ear."[20]

Tesla predicted that a thousand years hence the telephone and motion picture camera might be obsolete, but the principle of the rotating magnetic field would remain a vital, living thing, for all time to come.[21]

By 1931, although Tesla's news value had waned, he was not forgotten upon occasion. As a recipient of the Edison Medal, it was appropriate that he be invited by the press to pay tribute to Edison upon his death. On October 19, 1931, Tesla was quoted in his praise of Edison:

> He will occupy a unique and exalted position in the history of his native land which might well be proud of his great genius and undying achievements in the interest of humanity.[22]

Then Tesla added:

> He was by far the most successful and probably the last exponent of the purely empirical method of investigation. His mind was dominated by one idea—leave no stone unturned. . . . He had a veritable contempt for book learning and mathematical knowledge. . . . The recurrence of a phenomenon like Edison is not very likely. The profound change of conditions and the ever increasing necessity for theoretical training would seem to make it impossible.[23]

Even in eulogy, Tesla was unable to omit this subtle barb of criticism. Edison always affected him that way.

Other writers who had less reason to recall Edison's frailties chose to picture him as a lover of nature and animals, citing his familiarity with the wood creatures near his home. Did anyone remember the little mongrel dog and the stray animals purchased for twenty-five cents to prove the deadliness of alternating current?

The following year Tesla attended the funeral of his good friend, Behrend. With the death of Behrend went a part of Tesla's vitality. Close friends were growing fewer and the world was moving away. In his clean, but mended and pressed, cut-away, Tesla stood out as a figure from another day. His brown derby, too, distinguished him as a being from another era.

Only Kenneth Swezey and a few others were able to see through the out-moded garments to recognize the genius. Swezey, a radio research writer in 1924, had long admired Tesla. On their first meeting the young man was overwhelmed with a four-hour interview. They had dinner together at that first meeting—Swezey relishing a full meal while Tesla nibbled on romaine with parmesan cheese. The bill for $15.00 seemed exorbitant to the young science writer.

For many years thereafter the young disciple was plagued with three a.m. phone calls from Tesla which went on for hours with the younger man unable to get a word in.[24] The cost of hero worship was high but worth the cost. Disciples and admirers were growing inattentive, and reporters were inaccurate, so it was to the young Swezey that Tesla turned for the measure of respect so much needed and so greatly deserved, and it was the youngster who responded to the worth of the older man in a world where others who had been aware had either died or had forgotten.

Notes

1. John J. O'Neill, *Prodigal Genius,* (New York, N.Y.: Ives, Washburn, Inc., 1944), 230-231.
2. Ibid., 233-234.
3. Interview with Kenneth Swezey.
4. O'Neill, *Prodigal Genius,* 236.
5. Ibid., 311.
6. Nikola Tesla, "My Inventions," *Electrical Experimenter,* June 1919, 178.
7. Interview with Kenneth Swezey.
8. *Literary Digest,* September 1, 1917, 24.
9. Ibid., September 8, 1917, 25.
10. Helen B. Walters, *Nikola Tesla,* (New York, N.Y.: Thomas Y. Crowell Co., 1961), 147.
11. Tesla, "My Inventions," 178.
12. Microfilm letters, Johnson to Tesla, March 19, 1926. Library of Congress.
13. Laurence Schmeckebier, *Ivan Mestrovic, Sculptor and Patriot,* (New York, N.Y.: Syracuse University Press, 1959), 2.
14. Ibid., 1.
15. Hugo Gernsback, "Cold Fire," *Electrical Experimenter,* November 1919, 632.
16. John B. Kennedy, "When Woman Is Boss," *Colliers,* January 30, 1926, 17.
17. Ibid.
18. Alden Armagan, "A Famous Prophet of Science Looks Into the Future," *Popular Science Monthly,* November 1928, 16-17.
19. Ibid.
20. Ibid.
21. Ibid.
22. *New York Times,* October 19, 1931.
23. Ibid.
24. Interview with Kenneth Swezey.

Nikoa Tesla in his 60s. *Sarony studio photograph, from Kenneth M. Swezey Papers, Archives Center, National Museum of American History, Smithsonian Institution (AC0047-0000018).*

Tesla's Involvements with the Occult

From the moment when Tesla first reached national recognition, his life was no longer private. Every eccentricity became magnified as gossip was repeated and enlarged upon. True to human nature, the public was ready to cash in on the benefits of his super-intellect, but unable to understand that such excessive powers of concentration and perception might well take their toll in compensatory behavior patterns. The "herd" was willing to accept the fruits of genius as a matter of course, but was unable to grant any deviation from what was considered the norm. Freudian concepts were too new, and too shocking, for layman's credence and Victorian nicety prohibited any overt interest. The nonconformist was damned. As a result, Tesla's idiosyncrasies branded him as a mystic, charlatan, and crackpot. There were those who went so far as to maintain that he was a prophet whose origin was on another planet. Whatever the speculations, he remained an enigma.

His own extravagant statements were grist for the mill of every sensational journalist. Often when they could not understand either his advanced theories or his peculiarties, they made their own interpretations, which sometimes led to the destruction of Tesla's reputation as a rational scientist. He was misunderstood, misquoted, condemned—and lauded, to the complete confusion of the public. *The New York Sun* stated in 1896:

> Tesla has been the victim of more forged interviews and sensational articles appearing without authority than any other inventor.[1]

His most productive years chanced to coincide with an era of yellow journalism, industrial greed, and unlicensed piracy. A naive and gullible public watched the growth of fabulous fortunes ready to be gambled on any scheme for gain.

There was no denying that Tesla gave them all reason to talk. His early background was set in old world mystery. He came from a land criss-crossed by wandering gypsy bands with their gaudy wagons, smouldering campfires, furious dances, and the plaintive strains of violins. There, gypsy crones could decipher the future in palms crossed with silver. Like these wanderers from Romany, Tesla had lived at the crossroads of the east and west and was a blend of divergent civilizations.

Thomas Commerford Martin, once said of Tesla:

> He comes from an old Servian family whose members for centuries have kept watch and ward along the Turkish Frontier. . . . When a man springs from a people who have a hundred words for knife and only one for bread, it is a little unreasonable to urge him to be careful of his own life. . . .
>
> Mr. Tesla has been held a visionary, deceived by the flash of casual shooting stars; but the growing conviction of his professional brethren is that because he saw farther, he saw first the low light flickering on tangible new continents of science.[2]

Even as a child, Tesla's intense imagination had given a clue to his latent creativity—and to his difference from those around him. The visions which appeared before him, continually, became an affliction. These hallucinations may have been caused by some physical deviation of the optic nerve or by peculiar brain structure. Whatever the cause, they were frequent in times of danger or of great exhilaration. Tesla once described the phenomena:

> In some instances I have seen all the air around me filled with tongues of living flame. Their intensity, instead of diminishing, increased with time.
>
> . . . This luminous phenomena [sic] still manifest themselves from time to time as when a new idea opening up possibilities strikes me, but they are no

> longer exciting, being relatively of small intensity. When I close my eyes I invariably observe first a background of very dark uniform blue, not unlike the sky on a clear but starless night. In a few seconds this field becomes animated with innumerable scintillating flakes of green, arranged in several layers and advancing towards me. Then there appears to the right, a beautiful pattern of two systems of parallel and closely spaced lines at right angles to one another in all sorts of colors with yellow-green and gold predominating. Immediately thereafter, the lines grow brighter and the whole thing is thickly sprinkled with dots of twinkling light. This picture moves slowly across the field of vision and in about ten seconds vanishes to the left leaving behind a ground of rather unpleasant and inert gray which quickly gives way to a billowing sea of clouds seemingly trying to mould themselves in living shapes. It is curious that I cannot project a form into this gray until the second phase is relaxed. Every time, before falling to sleep, images of persons or objects flit before my view. When I see them, I know that I am about to lose consciousness. If they are absent and refuse to come it means a sleepless night.[3]

Tesla sometimes recalled strong winds on a mountain top and the exhilaration they produced upon him as a child when he felt that his body was light as cork and that he could leap and float in space for a long time. With maturity he concluded that this was a feat of his imagination rather than reality. But he never forgot his childish excitement and he relinquished the hallucinatory soarings with painful reluctance.

Such imaginations led to deep-seated phobias and compulsions which he could not forego and to which he became enslaved for his entire life. Such aberrations he described:

> . . . a violent aversion against the earrings of women but other ornaments, as bracelets, pleased me more or less according to the design. The sight of a pearl would almost give me a fit, but I was fascinated with the glitter

> of crystals or objects with plane surfaces. I would not touch the hair of other people except perhaps at the point of a revolver. I would get a fever by looking at a peach and if a piece of camphor was anywhere in the house, it caused me the keenest of discomfort. Even now I am not insensible to some of these upsetting impulses. When I drop little squares of paper in a dish filled with liquid, I always have a peculiar and awful taste in my mouth. I counted the steps in my walks, and calculated the cubic contents of soup plates, coffee cups and pieces of food—otherwise, my meal was unenjoyable. All repeated acts or operations I performed had to be divisible by three and if I missed I felt impelled to do it over again if it took hours.[4]

These peculiarities led naturally into an acute germ phobia, so intense that he was impelled to wash his hands for every reason and for no reason. His monogrammed white silk shirts took the brunt of constant laundry. His collars and handkerchiefs were discarded after only one use. He preferred to replace ties every week, at one dollar each, and the only colors he considered were red and black. His laboratory washroom was, of course, private.

Hotel head waiters had become accustomed to his demands. He preferred a table which no one else used. A fresh table cloth at every meal and two dozen clean napkins was his standing order. Although the silverware had been sterilized at his request, he was still compelled to clean each piece personally with a clean napkin before he could use it. He could not tolerate a fly, and if one should light upon his table, this would bring a demand for the removal of everything. Then the meal would have to be started all over again.

He had an aversion to shaking hands, although he did do so on occasion. He lived only in first-rate hotels where cleanliness was assured. In his laboratory his workers were expected to meet his own meticulous standards.[5]

In his early life he had made a blueprint of his direction and there were no paths of romance or of feminine involvements allowed to divert him from it. His mother and his sisters were the only women ever to approach his private life. He placed all women on a high pedestal beyond his reach. The field of invention received the whole pure and distilled energy of his personality. Any need for feminine companionship was solved by strict sublimation. By disciplining his thoughts, he was able to discipline his emotions. That his life was not complete no one knew better than he. Sometimes he spoke of the loneliness of his life. Actually, he had no regret, for he felt that no passion in the world was equal to the sublime moment when an inventor saw the satisfactory completion of his work. He was not antagonistic to women. Tesla enjoyed their company and paid them the utmost respect and courtliness. He was not opposed to marriage for people in other vocations; rather he believed that marriage might enrich the life and achievement of a writer or a musician, but he was certain that an inventor's mind would be diverted and that no woman could enter his life without lessening his potentialities.

He took no chances on any physical or emotional contact. Even the divine Sarah Bernhardt, who coyly dropped her handkerchief one time near his cafe table in Paris, failed to arouse more than courtesy in Tesla, who sprang to his feet to retrieve the bit of lace, avoided her eyes as he returned it, and immediately resumed his conversation with a friend, on the safe subject of wireless telegraphy.

Tesla was considerate of the women employees in his office. If they were asked to work overtime, he saw that they had a good dinner at Delmonico's and that they were transported by cab. He, himself, never rode with them, but followed at a discreet distance in another cab—and picked up the tab.

He had rigid ideas of women's fashions and appearance. He disliked fat women. He believed that the lines of a woman's dress should follow the lines of nature and not Paris dictates. At one time when his secretary appeared in a high style dress with a waist far below the normal line, Tesla considerately sent her home in a cab to change before anyone saw her in what he considered a monstrosity. He believed that women should not marry too young, "for then," he said, "men marry you for your beauty."

Nevertheless, he liked beautiful women and enjoyed the company of intellectual ones, but he always reserved the privilege of completely de-sexing them in his mind.

He furnished grist for the believers in occult theories by seeming, at times, to subscribe to a fraction of their beliefs—such as his conviction that amber was a magic substance having the properties of electrons. The Greeks had long before ascribed life to amber, and Tesla sought the scientific reason for this mystery. He also commented on the theory among some enlightened people, that the pearl was alive and that it grew more lustrous and beautiful in warm contact with the human body. This sensuous implication may have explained his revulsion to pearls.

He questioned whether the ancient mystics might not have had some actual understanding of a primitive kind of electricity. Tesla believed that Moses was undoubtedly a practical and skilled electrician far in advance of his time, and he was certain that the Ark had been constructed in such a manner that it became a condenser, that the friction of air against the curtains might have generated electricity that caused the light to appear at night. Therefore, it followed that the sons of Aaron who were killed on touching the Ark were the victims of high tension lightning discharge. Tesla was intrigued with the possibility of the scientific explanation of Biblical miracles. Further, he made wide use of Biblical symbols. The use of such symbols may well have had an obscure foundation in his subconscious indoctrination as a child in an intensely religious family. Could the qualities ascribed to the Holy Ghost, of power, energy, flame, have been persistent in his retentive mind? And could the symbol of the dove have struck a half-remembered chord?

Tesla felt, too, that the vestal fires of the Romans could have been electrical in their source. In a lecture printed in *The Manufacturer's Record*, September 9, 1915, Tesla said:

> The belt drive must have been known to engineers of that epoch [Roman Empire]. . . . Under favorable atmospheric conditions a belt may be transformed into a dynamic generator, capable of producing many

> striking actions. I have lighted incandescent lamps, operated motors, and performed numerous other equally interesting experiments with electricity drawn from belts and stored in tin cans.[6]

Such ventures into the realm of the mystic made him a target of the materialists and the realists.

The press seized upon his theories of planetary communications and his conviction that there might be life on other worlds. Newspapers stepped up circulation by printing sensational heads such as "TESLA MAY SIGNAL THE STARS." Such far-out statements gained him no friends among the conservatives. To men whose feet were solidly on the ground, there could be no tolerance for one whose head was in the clouds. When Tesla tried to excite the phlegmatic public with the vision of unlimited goals within his grasp, he said, "We can make all the artificial lightning we want and we will have no difficulty in making apparatus with a spark gap of a mile and it could be increased."[7] The uncomprehending public read, shrugged, and turned the page.

When Tesla said, "If there are intelligent inhabitants on Mars or any other planet, it seems we can do something to attract their attention," he was branded as crackpot, sacrilegious and perhaps dangerous to the foundations of faith.

As a result of Tesla's wild claims, two classes of the paranormal identified him as their own. On one side were the sincere seekers of truth beyond the veil. On the other, were those who sought to use his name only to give their own schemes a semblance of validity for financial gain or notoriety.

One of the most sensational examples of extravagant misrepresentation was the publication of the book *Can a Man Live Forever?* The author wrote a convincing account of a bill introduced and passed in Congress for the establishment of a National Institute of Science. Its purpose, he claimed, was to provide for the study of the human body with a view to the prolongation of life. He stated that two million dollars had been appropriated on January 17, 1902, when the bill was passed almost unanimously and he went on to give details of the operation of the Institute. After much consideration by the commissioners, Pueblo, Colorado, was chosen as the location on the 640 acres of land donated by the Colorado Coal and Iron Company on the shores of Lake Minnequa. Or so the story went.

The author claimed that Thomas Edison had predicted that the secret of life would be discovered within five years after the Institute went into operation. Seven departments were reputed to have been set up under the direction of actual, internationally known scientists. They were Natural Science, headed by Thomas Alva Edison, Psychology under the direction of S. Weir Mitchell, and such others as Physiology, Biology, Anatomy, and Anthropology—all under the leadership of prominent international figures. Colorado Governor Bell was said to have given the reception speech (printed on parchment and later placed in the cornerstone) at the opening of the Institute.

A News Bureau was to be maintained for the convenience of the press. Mr. Edison was reputed to have been on hand from the beginning to supervise personally the construction of the Natural Science building and to oversee the unpacking of cell mortars, glasses, instruments, and batteries. Every piece of his machinery was placed as he directed. Included in the directors of the Natural Science Institute were Professor W. K. Roentgen, of Germany, and Nickola (sic) Tesla of New York City.

The men selected for the life experiments were to be inmates of penitentiaries, under twenty-two years of age and of good health. The author claimed that the pulpits rang with praises of the project and suicides ceased as a result of the establishment of the Institute. Insanity became less common and prisons were being emptied because crime had been eliminated. The most dramatic incident cited was the alleged discovery by Edison and Dr. Johnson Dodson, head of the Physiology department, who, it was claimed, had discovered the secret of life. Vivian, the baby son of Dr. and Mrs. John Anson, was selected for experiment. The child was born on December 20, 1904. The cord was not tied as usual, but was cut about three inches from the body and drawn through a hole in a celluloid box and fastened. The child was to be fed a vital fluid through the cord which was to supplant the use of food. It was believed that eventually the vital organs would disappear, the elixir would be absorbed through the umbilical vein, the red corpuscles would be eliminated and only white ones remain.

A hospital chart listed the child's hourly progress. Later details showed that the baby began to talk at two months, was reading at the age of two years, and by his third birthday comprehended the story of his birth, knew the past history of the world, and realized that he was to be the forerunner of a new civilization. Vivian Anson, the author stated, grew to perfect manhood, gifted with infinite wisdom. His home on the Missouri River became the mecca of thousands.

The damning connection of this book with Tesla was that it made claims to an interview between Tesla and a woman reporter from Ann Arbor on December 26, 1904. The article was quoted as having appeared in *The Minneapolis Times* of that date.

The gist of the interview was that Tesla believed that the new elixir would mean perpetual life for the human race, that the nutriment would maintain perfect health, and that the new man would be a non-conductor of heat. In this fake interview, Tesla was quoted as stating that there was no waste in the earth and that it created its own necessary materials by the continual re-use of the same matter and was capable of drawing energy from the sun, moon, and stars. He supposedly said, "If there is a perfect entity in one, why not in the other. . . . God is infinite, and is working through the wisdom of mankind." He believed that Vivian Anson would be a model for the new race, that he would grow up to the most useful size for the utility of an organism such as his would be and then stop growing of his own volition. He was confident that the experiment would eliminate organic function, and that the boy would have no need of other food than the fluid, and would cease to breathe, and would not require sleep. The complete changeover of the blood structure from the replacement of the red corpuscles by the white, would be certain safeguard against infection. The news article gave the appearance of an authentic interview, despite the preposterous claims, except for one careless statement: "God is infinite."[8]

These three words were out of keeping with Tesla's belief at the time. Anyone familiar with his meat-machine conception of man would have been certain of fakery. But the majority of readers had only superficial knowledge of Tesla.

There was a second section of the book with a politically-tinged, subtly veiled, treatise on the future of mankind under radical ideologies. The book contained sketches of the world-eminent scientists purported to have staffed the Institute and Tesla's portrait was among them. Engravings of the buildings were generously sprinkled through the pages. There was a picture of a Utopian farm in 1904 made possible by the experiments at the Institute and of the political regime then said to be in power.

The careless reader might have overlooked the small detail that these events were represented as having occurred between the years of 1901 and 1924. Yet, a more careful scrutiny would reveal the publication date of 1898.

Furthermore, John C. Bell, listed as governor of the state in 1902, was never governor of Colorado, but rather was a state legislator from 1893 to 1903, and was designated as a Populist Democrat. Although the Populist party was in power in 1893–1895, with "Bloody Bridles" Davis Hanson Waite as governor, this was a radical minority group. They achieved power through advocacy of Free Silver, the vital issue of the day. At the time the book was published, Waite had already been put out by the conservatives. No doubt the Populists were smarting under defeat and unwilling to retreat. The book was an attempt to bolster Populist Party theories. The neat trick was to place the Institute story with its lurid fabrication in the front of the book as bait for the propaganda which followed as a second section.

Prophecy was a popular type of writing of the day. It had the same appeal that science fiction has today with the difference that *Can a Man Live Forever?* was not labeled fiction, but presented as documented gospel truth.

The facts are, there never was a National Institute of Science in Pueblo, nor did the great scientists whose authentic pictures appeared in the book have anything to do with such a fantastic project. Neither did Tesla give the interview. Anyone reading the book at the date of its publication would not have been deceived. To anyone stumbling on it by 1930, and not noting the copyright date, the convincing style of writing, the use of pictures and names would have at least suggested some vestige of validity to the overly credulous. In proof that this confusion is still evidenced, in 1962, a college librarian in one of the finest libraries in Colorado, was offered a rare edition of *Can a Man Live Forever?* It was represented by the book dealer as historical fact and priced accordingly!

The years did not lessen the abuse of Tesla's unique ideas. The promoters of cabalistic doctrines continued to put their own interpretation on his personality. Strange cults seized upon his every word and carried his ideas beyond any direction he had intended.

One such example was a book[9] dealing with mystics of the world with Tesla as the central figure. It claimed the origin of both Christ and Tesla to be on Venus and ascribed Tesla's birthplace to a space ship bound for earth. Tesla's mother was also credited as being a Venusian. His advent was seen as direct destiny for the purpose of bringing electrical knowledge and illumination to earth people. The dedication was to Nikola Tesla and the White Dove—his twin ray. The author believed that the Dove now shared with Tesla the etheric realms of the White Island of Shamballa over the Gobi Desert where they are still working for the benefit of mankind.

This author repeated and enlarged upon the strange story first told by John J. O'Neill of the great love which dominated Tesla's life—for a white dove with gray-tipped wings which came to him repeatedly. The story is that Tesla gave all the love he might have had for a woman to the dove, and that with its death came the end of his creative career. O'Neill, a science writer, was known to be interested in matters occult and was close enough to Tesla to have been told this intimate story.

Yet, another friend of Tesla who knew him for many years denied it as legend, saying it was purely imaginative writing. There were others who said the love of pigeons stemmed from the childhood fascination Tesla held for the homing pigeons of Smiljan. There were others who were convinced that he was trying to discover the secrets of radar through the study of the homing instinct of these birds.[10] This is quite possible.

Be the reason what it may, Tesla was a familiar figure on the New York streets in the later years of his life, where he fed the birds in Bryant Park and in front of St. Patrick's Cathedral. Even the New York papers commented from time to time on this daily routine. When he became too frail to care for the birds, he hired a red-haired Western Union boy to take over the twice-daily task. John Lucan, the messenger boy, saw to it that the pigeons were fed five pounds of corn at each feeding and that if there were any wounded ones, they received their share. He sometimes broke ice in the fountains to make sure they had drinking water.[11] The employees at hotels where Tesla lived were familiar with his concern for the birds. Sometimes they would bring him the sick or wounded ones picked up in the street. Hotel owners were not so broadminded about his turning his room into a bird sanctuary.

Those who knew him well were aware of the paradoxical character of the genius who could plan a death ray powerful enough to destroy 10,000 airplanes at a swoop—and in another mood could carefully spread towels on his window ledge and put out little cups of seed for his sick pigeons.

One young science writer who, perhaps, knew him better than any other person, admitted that he often walked along the streets with Tesla helping him to rescue unfortunate birds. This friend verified the story that Tesla could call them from their nests any time, day or night. There is no question that these birds had a vital meaning for him and were a part of his life.

O'Neill believed that Tesla saw some brilliant light in the eye of his favorite white bird—the one with the gray wing tips—and that with her death Tesla felt an even greater compulsion to care for the birds. They needed him and he needed to be needed. Even though Tesla claimed no comfort of belief in life after death, he toyed with the hope that his dove might return in the body of another.

Otis T. Carr claimed some experience with Tesla when he was asked to buy unsalted peanuts for the pigeons. Carr claimed to have been a daily visitor and Tesla's disciple for three years and claimed to have received scientific secrets from the inventor.

Carr also sharpened his poetry pencil and proclaimed in verse his own views of the atomic and space age. He protested that although man might split the atom, he would, in return, suffer frightful devastation. It might be possible to blast a man into space, he claimed, but he could not be brought down again. In this premature poetic prophecy, Carr proved that his imagination was not of Tesla's caliber. Some things he had not learned from Tesla. His limitations were showing and further, Carr had allied himself with the occult.

But no one fought harder than Tesla himself to disclaim any suggestion of mystic manifestations. He felt that such things only beclouded the issues so that fundamental truths which he wished to explain for the progress of mankind were likely to be buried under the mumbo-jumbo of the pseudo-supernatural. At least he knew that he dared not openly ally himself with those who delved in the mystical.

One instance of Tesla's adamant stand on such matters was revealed when representatives of the Ford Motor Company called on him one day to tell him of the formation of a society for the study of psychic phenomena. They asked him to join in the undertaking. This was a keen disappointment to Tesla, who had always hoped that Ford might someday need his engineering services. Tesla said, "I suppose those engineers never knew how close they came to being fired out of my office."

Later, schemers went so far as to use the prestige of Tesla's name to promote a flying saucer deal that had little existence except on paper. Such swindles did nothing for Tesla's reputation.

Certainly there are inexplicable experiences in Tesla's life which cannot be overlooked by the skeptic. How can one explain his phenomenal memory? What about his photographic and abnormal sense of sight and hearing? What of the admitted visions? What about the principle of the rotating magnetic field and his alternating current motor? How could he acquire his prophetic grasp of cosmic knowledge a century in advance? How does one rationalize the complete disappearance of certain documents, records, and data important to his life story?

For instance, the location of the Colorado Springs laboratory can be determined by the city directory, photographs showing other identifiable buildings, mountains in the background, landmarks, etc. But the plat books show no drawing of such a building, as they do of the surrounding ones. It is possible, of course, that ownership of the land might have been listed under another name. Newspaper records show that the building was finally disposed of and some equipment sold for taxes—there is no tax record of such transactions. A judgment was allowed against Tesla, according to news reports. The article lists the amount of the judgment, the name of the presiding judge, and the court in which it was tried—there are no records of such a court transaction in the County Clerk's office, nor in District Court files, nor the Denver Archives.

How does one account for Tesla's premonition of his mother's death? What was the significance of certain numbers which he respected—of strange augury he sensed on particular days? What of the coincidence of lightning with the auspicious moments in his life?

When Tesla died, obituaries appeared in the *New York Times*, but no notice of either his death or his funeral was ever printed in the block set aside for such information. Perhaps this was a matter of choice of the family—but why?

No photographs of the service appeared in the *New York Times*, although it was a Yugoslav state function and attended by national dignitaries and the scientific elite. A disciple of the occult explained that Tesla's body was not really inside the casket and that strange rays emanating from the casket had made photographs blurred and indistinct. But Kenneth Swezey, the good friend of Tesla, produced one picture which was clear in every respect. He explained that the slower speed of films at that time and the necessity to do with available light made many of the pictures indistinct when no tripod was used. He refuted a mystical explanation of the picture story.

No one will ever be able to interpret Tesla, although every biographer tries. J. J. O'Neill probably made the most plausible explanation—that Tesla was a man in the advanced state of development, that the combination of this and a superhuman drive and unflagging work made the right formula for his success. But what was the secret of the superhuman drive? And what causes a being to be born out of his time?

Even a blurred photograph taken at Nikola Tesla's funeral fueled theories that mystical forces were at work. Cathedral of St. John The Divine, New York, January 12, 1943. *From Kenneth M. Swezey, Courtesy of Fran Kiester.*

The meaning of "occult" is varied and subject to individual interpretation. For some it bears the connotation of the supernatural—the spell of witchcraft—the lure of astrology—and the tinge of black magic. To others the term simply implies that which is hidden and yet to be discovered. To the latter group, all knowledge is occult until the period of discovery. Electricity is a perfect example—it has always existed, but was occult or hidden until science discovered its laws. There are still many facets of electricity which are harnessed, yet not entirely understood. So the veil between those who identify themselves with explorers of the occult, and those who protest, is often thinner than either group wishes to acknowledge. One well-known science writer says, "It is always the far-out dreamer, damned by the ones who insist on laws and logic, who has pointed the way to progress."

One writer ventures:

> Out of the sunset, came Tesla's greatest invention and his career was launched. Out of the dazzling white light from a pigeon's eye, came that career's end.[12]

Today's concept of the universe is unlimited. There are no longer any absolutes. The acceleration of new discoveries has brought about confusion which thins the hair-line between the scientific and the occult. One can only conjecture, "Was Nikola Tesla a mystic?"

Notes

1. *Electrical Review*, November 30, 1898, 344.

2. Thomas Commerford Martin, "Nikola Tesla," *Century*, February 1894, 582-585.

3. Nikola Tesla, "My Inventions," *Electrical Experimenter*, February 1919, 745.

4. Ibid., 745-756.

5. John J. O'Neill, *Prodigal Genius*, (New York, N.Y.: Ives, Washburn, Inc., 1944), 289-295.

6. *Nikola Tesla, 1856 – 1943, Lectures, Patents, and Articles*, (Beograd, Yugoslavia: Nikola Tesla Museum, 1956), A-177.

7. "Is Tesla to Signal the Stars?" in *Electrical World*, April 4, 1896, 369.

8. J. Emile Hix, *Can A Man Live Forever?*, (Chicago, Ill.: The Western News Co. Publishers and Agents, 1898).

9. Margaret Storm, *Return of the Dove*, (Baltimore, Md.: A Margaret Storm Publication, 1959).

10. Interview with Frank Waters.

11. *New York Times*, May 1, 1937, 21.

12. Gaston Burridge, "Was Nikola Tesla a Mystic?," *Rosecrucian Digest*, January 1956, 36.

An incandescent lamp is lighted by current passing from the large "extra coil" in Tesla's Colorado Springs laboratory to "the structure of iron pipes," according to Tesla's description of this photograph in his notes. Tesla built the "extra coil" on a wooden frame eight feet in diameter and it was used "chiefly for investigating propagation of waves through the ground and similar objects" and to "investigate further the behavior of strong streamers." *Courtesy Nikola Tesla Museum, Belgrade, Serbia, (006-02-052).*

Robert Underwood Johnson. *Harris & Ewing photograph, Library of Congress Prints and Photographs Division Washington, D.C., (LC-DIG-hec-19685).*

Tesla's Friendship with Robert Underwood Johnson

Although Tesla seemed to live an almost solitary existence, he did treasure a very few close friends. His relationship with those few revealed certain unique facets of his personality and offered clues to an understanding of this unusual man. Because he could only claim a hotel room as residence, he warmed himself at the fireside of the Robert Underwood Johnsons on Lexington Avenue. This family association became the nearest approach to a home that he ever knew. It was at the Johnsons that he met friends of a caliber which he appreciated.

Tesla was introduced to Robert Underwood Johnson in 1893 by Thomas Commerford Martin, later president of the American Institute of Electrical Engineers. This was the beginning of the life-long friendship between the two. The friendly circle expanded to include such notables as Mark Twain, Joseph Jefferson, John Muir, Captain Richmond Pearson Hobson, Gericke (the conductor of the Boston Symphony Orchestra), Madame Milka Ternina (the great prima donna), Maurice de Monvel (the French painter), and Ignace Paderewski.

Here in the gracious home of the Johnsons came the cultural aristocracy of the day. Johnson, as editor of *Century*, drew the literary elite. Such people as Helen Hunt Jackson, Mrs. Vincent Botta, Thomas Wentworth Higginson, and the Hearsts were regular visitors.

Johnson and Tesla shared a common interest in cultural arts but both men were also vitally interested in scientific developments. At the age of eleven Johnson had assisted a station agent near his Indiana home as telegrapher. He had become an expert at sending dispatches and he recalled listening for the "autograph" of an operator at the Indianapolis office who signed with a special touch, "Edison."

Even as a young lad, Johnson recognized greatness, and when he took the message of Lincoln's assassination over the wires, he decided that he must follow the funeral procession to Indianapolis. He was so deeply moved by the solemnity and the dignity of death and the half-realized historic significance of the occasion that he went through the silent procession a second time.[1] There was no morbid curiosity in his compulsion to do this. It was only that he had been deeply moved with the innate sensitivity of a poet.

Earlham College, a Quaker school at Richmond, Indiana, admitted Johnson when he was only fourteen years of age, and granted him his Bachelor of Science degree when he was eighteen.

He went to work for *Scribner's Monthly* in Chicago and then in New York, as associate editor, where the periodical later became *Century Magazine*. In 1879 when Scribner's published the first authoritative account of Edison's electric light, Johnson visited the inventor's laboratory at Menlo Park. He was keyed up to concert pitch as was every one at the laboratory. "Edison, the great inventor," made almost as intense impression upon him as had the "telegraph operator, Edison," on the "night line" in Indianapolis. Young Johnson was given an opportunity to buy stock in a chance to light the world, but turned down the offer, a decision he lived to regret. Brilliant as he was, he had a gift for betting on the wrong horse.

As staff member of *Century*, he had opportunity to widen his interests. The popular writers of the day were contributors to his magazine. Several of these authors were anonymous and their real names were known only to Richard Watson Gilder, editor in chief, and in one instance, Thomas Wentworth Higginson. Higginson's acquaintance with Helen Hunt Jackson and Emily Dickinson let him in on the secret of Helen Hunt Jackson's "Saxe Holm" stories. Johnson, himself, began to dabble in creative writing. Poetry was becoming his interest. He was equally competent in the judgment of scientific writing. Naturally he and Tesla were drawn together by these common interests.

Johnson realized that Tesla's inventions were not in the same category as those of Bell or Edison and that few people would ever grasp the full magnitude of Tesla's discoveries of the principles of the rotating magnetic field which bid sure to revolutionize the world of industry. At the time the two men became acquainted, Tesla was engaged in some of his most profound experiments. The rapport between them was instantaneous.

Before the Johnsons' fireside, Tesla relaxed and revealed his wildest prophecies of airplanes to be operated by wireless, of controlled torpedo boats, streetcars running in London by Niagara power, and communication with friends at any distance by means of a pocket instrument and a wire stuck in the ground. Johnson believed implicitly in his friend. Sometimes other guests shrugged their shoulders and tapped their heads meaningfully. Johnson recognized Tesla as the most imaginative of all electrical inventors. He grieved to see how others pilfered Tesla's ideas and picked his brains to reap where they had not sown.

At the Johnson table was the first meeting of Tesla and Ignace Jan Paderewski. They became congenial friends at once. The two men shared a deep love of music and an intense interest in European politics. Johnson declared, "Two more intellectual and lovable men, I have never known."[2] Antonin Dvorak was another composer-musician who enjoyed companionship with Tesla at the Johnsons.

One of the most welcome introductions for Tesla was to the intensive guest, Mark Twain, with his strange contrast of traits. "One of the hardest hitters and hardest haters in all literature," said Johnson.[3] Tesla recalled his youth in Gospich when he had read avidly Twain's *Connecticut Yankee in King Arthur's Court*. He marveled at the strange thread of incidence which might be more than coincidence. Twain was one of those who enjoyed going to Tesla's laboratory with other Johnson guests to witness brilliant electrical experiments. When Tesla invented the first phosphorescent light for indoor photography, Twain and Johnson were among those who were given photographs as souvenirs of their visit. On one such evening, Johnson proved his faith in Tesla's skill by allowing a current of a million volts to be passed through his body to light the lamps in the room.

The popular English author, Rudyard Kipling, was another guest invited to the Johnsons with Tesla and other notables. Both Johnson and Tesla admired the vigorous personality of

Kipling. When the latter succumbed to typhoid and was unable to come to the dinner, the Johnsons notified all the guests and called off the party. Tesla was quick to express concern over Kipling. The only guest who failed to receive notice of change in plans was Mrs. George [Elizabeth Bacon] Custer, who, unaware of Kipling's illness, showed up "in her best gown and smile." The Johnsons graciously invited her to be the lone guest, and after dinner she rewarded their hospitality with a graphic account of her life in the far west with the general, where she was the only woman who rode with the cavalry. She described both the grim and the ludicrous experiences of the army wives who often lived on the outposts "eighty miles from a lemon."

Mark Twain in Tesla's laboratory, experimenting with fluorescent lamps for photography. *From* Remembered Yesterdays, *by Robert Underwood Johnson (Little, Brown & Co.), courtesy of Fran Kiester.*

Tesla and Johnson were closely bound by their love of poetry. Tesla could be given any line from a part of Faust and could pick up the quotation and continue by memory, page by page. He could recite long passages of Serbian epics, and he made literal translations for Johnson which became the basis of the "Paraphrases" from the Serbian poet, Zmai Iovan Iovanovich. In their discussion of the poetry of the Balkans, Tesla translated the poem about the legendary hero, Luka Filipov. It was a moment of intimacy in the Johnson household when

Tesla gave the affectionate nickname of Luka Filipov to Johnson. From that moment, Johnson adopted the signature for their friendly letters and greetings, and Mrs. Johnson became Madame Filipov. These were the only friends with whom Tesla was ever known to use or tolerate such informality of address.

Luka Filipov, hero of the Montenegrin War of 1874. *From* Century Magazine *courtesy of Fran Kiester.*

The children of the Johnsons were also included in this inner circle, and when Tesla would arrive by cab and extravagantly order the driver to wait during the visit, the children were allowed to ride through the park in his hired rig.

Not all the Johnsons' guests were aware of Tesla's importance. Johnson was amused one evening when he heard a feminine English writer ask, "And you, Mr. Tesla, what do you do?" The quiet reply was, "Oh, I dabble a little in electricity." The lady smiled condescendingly and said, "Indeed! Keep at it Mr. Tesla and don't be discouraged. You may end by doing something some day." "This," said Johnson, "to the man who had sold the invention used at Niagara Falls to the Westinghouse Company for a million dollars and lived to rue the bargain."[4]

The friendship between the Johnson family and Tesla became more firmly cemented with the years. Family joys and sorrows were shared. The two men corresponded almost daily at times with formal notes of invitation and inquiry. Their letters revealed a warmth and wit seldom suspected by the outside world. One such letter was an invitation to the Johnsons for dinner at Delmonico's and theatre afterward. Tesla expressed his regret at having to offer them seats no nearer than the fifteenth row which he felt "would necessitate the use of telescopes but might be better for Mrs. Johnson's imagination."[5]

Though the Johnsons were not wealthy, they were accepted by the well-to-do and the most aristocratic friends. Sometimes the two men joked about their fashionable acquaintances. "When millionaires desert you," wrote Johnson, "our humble hearth of poverty will always have a welcome for you."[6]

"Dear Luka," Tesla often began his letters. Sometimes he would close on a light note, signing, "N. Tesla, G. I. (Great Inventor)" or again, he would sign, Nikolas I (of Houston Street.)[7]

He kept in close touch with the Johnsons when he was in Colorado. Perhaps they both felt a sympathetic bond in their impracticability. Neither of them knew how to make or keep a dollar consistently. Both men had known the pain of rejection. Tesla wrote:

> Luka, I see every day that we are both ahead of our time. My system of wireless telegraphy is in the transactions of a scientific society and your great poems on the heroes of Manila did not even as much as save Montejo . . . but we shall continue in our noble efforts, my friend, not misregarding (sic) the bad and foolish world and sometime when the fellows who form syndicates and write poems for Rogers, Peer and Co. shall have furnished material for feats of unspeakable destruction, I shall be explaining the principles of my intelligent machine (which will have done away with guns and battleships) to Archimedes and you will read your great poems to Homer, but I have no doubt that the old gentleman (who I hope has mastered the English by this time) will tell you that they are the finest ever heard. . . . The contemporaries will never understand us. We have an impediment of speech when it comes to their language.[8]

Referring to his Colorado experiment, he wrote:

> . . . I have had a wonderful experience here. Among other things I have tamed a wild cat and I am nothing but a mass of bleeding scratches, but in the scratches, Luka, there lives a mind.—Mind! . . .[9]

With his usual warmth and humor he once closed a letter with an apology that he had not yet carried out his promise to become a millionaire. "Kind regards to all of 327 Lexington Avenue," he said, "right close to the streetcar line—behind the big pole on the street."[10]

Johnson, as editor of *Century*, seldom published a scientific article without conferring with Tesla. One such occasion brought a plea to Tesla to scan the proposed article for error, asking him to be frank—*"As thou lovest me!"*

When the Johnsons sought relief from New York summer heat, at their cottage on the New England coast, it was with a feeling of guilt they remembered their friend in his New York laboratory. Once Johnson wrote, urging him to join them but with little anticipation that the invitation would be accepted, since, as he wrote, he was convinced that Tesla was "joined to his idols of copper and steel and would not think of getting more than three miles away from Delmonico's."[11]

Another summer when the Johnsons had taken refuge at York Harbor, Maine, they again urged Tesla to join them. They wrote:

> General Sherman's son who arrived from New York says the temperature is what his father said war is. But cheer up: we have discovered a way of escape for you. Six miles north of here on a bluff above the sea is a beautiful first class hotel where we can have the best of food and drink, cooked to equal the Delmonidor-Astoria with suites including private bath . . . common people must have rest like machinery but the great old Nick—the Busy One—is a devil of a fellow. See him go—150 hours without food or drink.
>
> Why, he can invent with his hand tied behind his back! He can do anything in short, but be sensible

> and prudent. He is superior to all laws of hygiene and human energy. He is a vegetarian who doesn't know how to vegetate! Down with the pretender. Abas Tesla. Hurrah for Pupin the First! Viva Moore! Viva Tripler! They know a summer resort when they see one at least. They know enough to come in when it reigns.[12]

Such letters brought laughter to Tesla but he felt no need for vacation. Once Johnson had pleaded, "Be good to yourself. If you don't come you are a ________." And in the blank, Johnson sketched a chicken![13]

Often Johnson requested an article for *Century* from Tesla. In such instances he did not hesitate to return any writing for revision. He was unwilling to publish an ineffectual article by his best friend. "Trust my thirty years of experience in knowing what the public finds interesting," he begged Tesla. "Faithful are the wounds of a friend."[14]

Often, Tesla poured out his irritations to Johnson. After his famous 1900 *Century* article, when he was attacked by an army officer as being subversively in favor of doing away with armed forces, it was to Johnson he turned to explain that he realized that it was only in an ideal world that military force could be dispensed with, although he felt that it constituted a state contrary to the teaching of the Bible.[15]

Both men turned instinctively to each other in times of frustration and attack. At one time when Johnson was smarting under the critical lance of a reviewer of his *Paraphrases,* he complained to Tesla that the reviewer had failed to cut the last twenty pages of the book. He could count on Tesla for a sympathetic ear. The two men treasured their friendship.

When Wardenclyffe Tower appeared to be doomed for lack of funds, it was Johnson on whom Tesla called to look over his plans for a manifesto, advertising his availability as a consultant engineer. It was to Johnson he wrote for a millionaire mailing list and to him he submitted the first draft for correction of his grammar in the Latin proverb which he used to head the manifesto. (*Nihil In Sacculo Quod Non Fuerit In Capite.*) This was a favorite proverb. Translated, it meant, "There is nothing in the pocket that was not first in the head." In this period of bitter disappointment he wrote to Johnson:

> I hope the Filipovs are happy to the extent that I am unhappy. That would be bliss.[16]

It was natural that Tesla and Johnson should rely on each other for financial help at times. Johnson once deposited $2,015 to Tesla's account for which he was given twenty-five shares in the Nikola Tesla Company for security. Tesla kept accurate account of all these transactions.

As the years progressed, both men found their bank accounts dwindling. Adversity came to both. Neither man had a particle of business acumen or practicability. Tesla's money was spun out for copper wire and the Johnson family took off for Europe on the slightest pretext. Both men sensed impending change in the economic picture of the country but neither had the ability to shore up his resources or cut back expenditures. The Johnson house, built in an era of extravagance, grew far too expensive to maintain and the bills for entertainment reached a dangerous high but there simply was no other way of life.

Johnson's position as editor of *Century* grew precarious in 1913. His letters to Tesla became mysterious, veiled and secretive. He urged Tesla to keep the correspondence confidential. Evidently Tesla destroyed certain letters, for there is a gap in their story. On June 2, 1913, the *Century* Company announced Johnson's resignation. Probably it was under some pressure.

Years before, the American Academy of Arts and Letters had had its inception in the *Century* office of its editor, Gilder, and Johnson had held an office in the organization since 1905. It was to that foundation he turned in his troubled year of 1914 for work. Problems seemed to be mushrooming. His brokers were about to sell his copper stock because he could not raise the margin. The financial world was aware of his straightened circumstances, but the social whirl continued to include him in its expensive orbit. There was no courteous way of escape.

Tesla tried to bolster Johnson in his time of depression, assuring him that soon he would be able to assume all financial burdens for both of them. He wrote:

> Do not worry about finances. Remember while you sleep, I work and am solving your problems. The old clippings you have forwarded are a sad register of my former folly. I had 36 patents on my system of power transmission in which billions are invested now. I won every suit without exception and had it not been for a "scrap of paper" I would have received royalties, Rockefeller's fortune. But just the same I feel I am safe to invite you to dinner for the New Year after next in my own Town House. Need I say that we shall have a fine time in the summer at my country residence?
>
> P.S. Look to a formal notice from my bankers that you may draw on me for all you want.[17]

So he brought a smile and shared an aircastle with Johnson. A few days later he sent another such letter:

> Write your splendid poetry in serenity. I will do away with all your worries. Your talent cannot be turned into money, thanks to the lack of discernment of the people in this country, but mine is one that can be turned into carloads of gold. I am doing this now.[18]

Johnson answered, "Wordsworth said, 'Poetry is emotion remembered in tranquility.' The problem is how to get the tranquility."[19]

Creditors continued to nag Johnson. Tesla found himself embroiled in a law suit against Guglielmo Marconi. Both men leaned heavily upon Fate to solve their problems. Despairing of God, who seemed to have turned His face at the moment, Johnson implored Tesla to pray for him to the "No-god" that he worshipped.[20] The stars of both men were diminishing.

Tesla was among the guests invited to the Johnson home for Christmas dinner in 1915, but he declined. The Nobel

Prize fiasco may have had something to do with it. Perhaps he could not bear to discuss it. The shock may have been too great. Again he was urged to drop in on the Johnsons for a birthday dinner. "On your way home to your drear lodgings at the Waldorf," Johnson had said, but Tesla declined.

Guglielmo Marconi, ca. 1903, was credited for the first transatlantic radio transmission in December 1901. He was issued a U.S. patent for an "apparatus for wireless telegraphy"—the radio—in June 1904. In June 1943, the U.S. Supreme Court decided that Nikola Tesla was the true inventor of radio. *Library of Congress Prints and Photographs Division Washington, D.C., (LC-USZ62-121714).*

The following year, Tesla received the customary invitations to spend Thanksgiving and Christ-mas holidays with his friend, and the letters from Johnson indicated an impending operation and precarious health. He begged Tesla to spend time with him then, rather than write an elegy for him later.[21]

Johnson's ill health brought about a certain petulancy and change of temperament. He wrote to remind Tesla of $2,000 which he had once borrowed. Tesla immediately mailed a check for $500, despite his own financial needs. In about two weeks he received another letter from Johnson enumerating his pressing bills for taxes and interest amounting to $967 and asking if he could carry out his intentions and repay more of the old debt. Tesla sent another $500 check. Before the end of the year, Johnson reported a bank balance of only $19.41 and debts of $1,500. He begged:

> If you can send me $500 I should invoke the blessings of Mercury, god of commerce, on whom your fortunes during the year of 1919 depend, which everybody says

is to be so prosperous. I have nothing in sight that I do not already owe.[22]

Tesla scraped up the check.

Johnson took hope in a rumor, by March, 1919, that he might be appointed Ambassador to Italy for the United States. He wrote a friendly letter to Tesla informing him that Marconi was now talking of communication with planets and that the young inventor was not being ridiculed as Tesla had been. "Moral," Johnson pointed out: "Better be second than first in some things." Then he added, "Wouldn't it be strange if that automobile you always promised me might have to be delivered to the Eternal City?" "Meanwhile," he concluded, "walking continues in this neighborhood."[23]

"The newly appointed Ambassador to Italy, Robert Underwood Johnson photographed as he left the White House after his interview with the President." *National Photo Co. photograph, Library of Congress Prints and Photographs Division Washington, D.C., (LC-DIG-npcc-01177).*

But Johnson's gay spirits were brief. Interest on the mortgage and penalty on his taxes came due and dire necessity induced him to go to Tesla again. This time Tesla found himself unable to help. "If you can collect $2,000 from Richard Peters who owes me, you can have it," Tesla replied wearily.[24]

Johnson decided to suggest a few money making schemes to Tesla to see if he could prod him in the right direction for both of them. He began to suggest the possibility of reproducing football games on screens simultaneously with the action. He urged Tesla to play with the idea. Tesla replied cryptically:

I am already expecting to become a multimillionaire without going into the show business, but if your ambitions are in that direction I shall assist you. . . . best suggestion is to employ nine flying machines, winged and propellerless five hundred miles or more, take negatives, develop films, and reel them off as they arrive. I presume you do not care for expense. It calls for an invention to which I have devoted twenty years of careful study, which I hope will ultimately realize, that is, television, making possible to see at distance through a wire. I agree with you, but if successfully carried out, it could be adapted for things which you do not seem to have thought. I am glad, however, to know that you are a genuine American, "*baseball uber alles*."[25]

Katharine McMahon Johnson, "Madame Filipov," married Robert in 1876. Their friendship with Nikola Tesla was the closest he knew during his life in the U.S. *From* Remembered Yesterdays, *by Robert Underwood Johnson (Little, Brown & Co.), courtesy of Fran Kiester.*

In addition to financial worries, bad health became a spectre in the Johnson household. Madame Filipov was ill and Tesla was quick to prescribe dietary precautions. He had a feeling of apprehension about her. A break in the circle of the only family he knew was unthinkable. Many women had admired him, but Madame Filipov was the only one who had ever been truly concerned about him.

Johnson shared with Tesla the joy of receiving an honor from the Order of St. Sava by the Prince Regent of Yugoslavia. It had been a recognition of Johnson's Serbian poetry. Tesla congratulated Johnson, but added that for himself, his American citizenship papers denoted an honor prized above all others.[26]

"Ought you not to value your birth certificate more?" quipped Johnson. "Where would you be without that?" "But," he added seriously, "I, too, would rather write a good poem than receive an honor." Later that month, he cashed in a few government bonds to have some more poems printed by the Yale University Press.

At last the information that Johnson was to go to Rome came through, but not until after he had already accepted a position with the Hall of Fame of Great Americans which he had to postpone. He was busy winding up his affairs but not too busy to rise to the defense of Tesla when Charles Proteus Steinmetz took the spotlight in the news by repeating and receiving credit for experiments which Johnson felt Tesla had done years before. Johnson also begged Tesla to look up some old pictures of Mark Twain and others who had posed in the laboratory for photography with fluorescent light. Johnson was beginning to write his memoirs. Tesla would be a part of his story.

In 1924 he returned from Italy and resumed the position he had postponed with the Hall of Fame. For over a year the shadow of death had seemed to hover over the Johnson household. It had begun with the death of the beloved daughter-in-law, a heartbreak experience to the entire family.

The following year, little Madame Filipov, also, died. Both Tesla and Johnson were stricken with grief for this beautiful woman whom they had both cherished. Her dying request was for Johnson to keep in close touch with Tesla.

Johnson tried to carry on tradition by asking Tesla to join him at a family dinner on her birthday. "We will have music," he said, "the kind of occasion she would have desired. She cherished your friendship. She charged me not to lose sight of you. Without you it will not be her day."[27]

By June of that year, Johnson was in a financial bind again with income tax and a note due at the bank and changes in his expensive old house which were demanded by the city building code. He again pressed his friend for help. Tesla reassured him. "Do not let those small troubles worry you," he said. "Just a little longer and you will be able to indulge in flights on your Pegasus."[28] Tesla was recovering from an illness but he managed to scrape up enough to tide Johnson over the difficulty. There were some engineering jobs—a few royalties, an occasional new invention, and sometimes a donation from a financier which could be kited to help them both over the rough spots.

Johnson wrote promptly to say, "Bless you for your enclosure." Tesla's check must have lightened the situation substantially, for Johnson added, "Agnes [his daughter] and I are going to sail for a two months trip to the low countries, finding it cheaper to close the house and pay the servants half wages. A happy summer to you."[29]

When Johnson returned to New York, he immediately wrote to Tesla reminding him of Madame Filipov's dying request for their enduring friendship. There was again to be holiday gaiety at the Johnson home. In special consideration of Tesla, he hired a Serbian musician to play for the party.

With a sudden wave of good luck, Tesla sent an unasked-for check for $500 to Johnson urging him to have some relaxation on the occasion of Madame Filipov's birthday. Johnson took half the check and had an inscription put on the wall at the foot of her grave.

By August, income tax and expenses were again forgotten and Johnson took off for Europe again, writing from Switzerland to say that he had seen Tesla's old flame, Venus de Milo, at the Louvre.

Returning home, he invited Tesla to unveil the bust of Benjamin Franklin in the Hall of Fame. He reminded him that Edison had already done the same thing for Joseph Henry.[30] Tesla refused the honor. He cared little for the publicity it might have given him. He preferred to walk in Bryant Park and in front of St. Patrick's Cathedral and coax the pigeons to eat. Johnson wrote, "My love to your pigeons."

By May of that year Johnson was off again to Europe. In his letters, he referred to Lindbergh and the great things the lad was doing as a sequel to his flying achievement. He chided Tesla for not writing. He quoted Goldsmith:

> He threw off his friends as a hunter his pack. He knew when he wished he could whistle them back.[31]

Early in the next year it was Tesla who was feeling the pinch for money. Johnson sent him a check for $250 with a curt reminder that it was not a gift and that he needed it back.[32] The following day Tesla returned the money. He said he had had a fainting spell, and it was a case of the lion saving the mouse. In two weeks it was Johnson who was begging to be saved from an overdrawn bank account.

Johnson had been invited to go to Europe to the Ibsen Centenary, all expenses paid, but he preferred to go with a free mind, and so he refused expense money and was cramped budget-wise. The result was that by June he was pressing Tesla for more help. But within a month, he flitted off to Europe again for a real rest and for the Centenary—with a free mind.

Tesla plugged on in his laboratory on a delicate invention that a good press agent would have had the world talking about. He felt that his prospects were better than ever. He wished that his fainting spells would not come so frequently. Johnson returned home to become concerned over Tesla's health. He questioned him to find out who would know about his affairs if something should happen. Tesla shrugged off the possibility of illness and Johnson scarcely noticed, for he was already concentrating on a trip to Ireland. He sent a picture postcard of the Shannon power machinery at Limerick to Tesla. He noted that it was the same installation as used at Niagara Falls.

He returned in time to go to Madame Filipov's grave on the anniversary of their wedding. A postcard to Tesla reminded him of the magic circle of three that had been broken. It was a grief they would always share.

In 1931 Tesla had moved to the Governor Clinton and had returned a loan to Johnson, saying thanks would come later by freight. "My loafing days are over," he said, "I am working now and hoping you see the world as if it were made for you."

Tesla began to haunt the movies for relaxation. This seemed out of character for a man with his intellectual capacity. Perhaps it was the refuge of darkness. Could it have been to study some inventive idea about the movie industry? Or was it sufficiently un-stimulating to be restful? Or was it the eternal lure of a story?

Johnson received a favorable review of his poems in the *New York Times* which warmed his heart. The reviewer ventured that Johnson might well be called the "poet laureate of America." But his pleasure was short lived.

In July, 1933, Tesla received communication that Johnson had suffered a heart attack in Stockbridge. The letter had been written for him but he had printed his name in crayon with childish letters. There was a senile indication in the reminiscent quality of the message. It was April of the next year before Johnson was able to return to New York. In July, Tesla again received a letter telling of another attack. An infection of the bladder was added to the complication of the heart condition. Johnson remembered that Tesla would be seventy-nine the following Wednesday and asked Tesla for a magic cure from his doctor. He said, "I have published a book at eighty-three called *Your Hall of Fame* but I shall not live to see your bust placed there, with impressive ceremonies. But there it will be, never doubt, my great and good friend!" He mentioned a recent recognition which had been accorded to Tesla in Belgrade and begged his old friend to give the *Times* editor all the details.

Johnson lingered in ill health until 1937. His last letter to Tesla was dated April 25th, of that year. It was signed with a trembling hand. Johnson died on October 14, 1937. His only legacy to Tesla consisted of a lifetime of memories and a handful of poems. There had been one poem written especially for Tesla. It was titled:

IN TESLA'S LABORATORY

Now in the dark what ghostly figures press!
No phantom of the Past, or grim or sad,
No wailing spirit of woe; no spectre clad
In white and wandering cloud whose dumb distress
Is that its crime it never may confess;
No shape from the strewn sea; nor they that add
The link of Life and Death—the tearless mad;
That live nor die in dreary nothingness:

But blessed spirits waiting to be born—
Thoughts to unlock the fettering chains of Things;
The Better Time; the Universal Good.
Their smile is like the joyous break of morn;
How fair, how near, how wistfully they brood!
Listen, that murmur is of angels' wings.

So the second link in the magic circle that had been perfect in trinity was broken. Tesla, alone, remained with an engulfing sense of aloneness. Changed as Robert Underwood Johnson had become in the latter years of his life, he had still remained a loyal friend. With his death there was no one left to carry out Madame Filipov's admonition to "keep in touch with Tesla." For Tesla there was no comforting assurance of immortality—only the empty ache of loss.

Perhaps, most of all, in this wounding moment, he missed the gentleness of Madame Filipov, who had long ago vanished from the scene. Vanished as a white dove might disappear, winging into the obscurity of a white cloud.

Notes

1. Robert Underwood Johnson, *Remembered Yesterdays,* (New York, N.Y.: Little, Brown & Co., 1923).
2. Ibid.
3. Ibid.
4. Ibid.
5. Microfilm letters, Tesla to Johnson, December 7, 1893, Library of Congress.
6. Ibid., Johnson to Tesla, December 28, 1897.
7. Ibid., Tesla to Johnson, September 29, 1898.
8. Ibid., October 1, 1899.
9. Ibid.
10. Ibid.
11. Ibid., Johnson to Tesla, July 28, 1896.
12. Ibid., July 17, 1900.
13. Ibid., August 24, 1901.
14. Ibid., March 6, 1900.
15. Ibid., Tesla to Johnson, June 22, 1900.
16. Ibid., December 15, 1903.
17. Ibid., December 24, 1914.

18. Ibid., December 27, 1914.
19. Ibid., Johnson to Tesla, December 26, 1914.
20. Ibid., September 13, 1915.
21. Ibid., December 21, 1917.
22. Ibid., December 30, 1919.
23. Ibid., March 19, 1919.
24. Ibid., Tesla to Johnson, April 11, 1919.
25. Ibid., November 29, 1919.
26. Ibid., January 6, 1920.
27. Ibid., Johnson to Tesla, April 9, 1925.
28. Ibid., Tesla to Johnson, June 3, 1925.
29. Ibid., Johnson to Tesla, June 24, 1925.
30. Ibid., February 17, 1927.
31. Ibid., undated letter.
32. Ibid., January 5, 1928.

Nikola Tesla (age 77), October 13, 1933. *Sarony studio photograph from Kenneth M. Swezey Papers, Archives Center, National Museum of American History, Smithsonian Institution (AC0047-0000019).*

Cumulative Honors Awarded to Tesla

Six years before the death of Robert Underwood Johnson, a flurry of honors came to brighten Tesla's days. The world-wide enthusiastic plaudits came about through the efforts of Kenneth Swezey, the young science writer. This disciple sensed the loneness of the great genius and deplored a world that was fast forgetting his major contributions. There was little Swezey could do materially to right the myriad of wrongs or to restore the well-deserved glory, but the younger man conceived the idea of searching out the great of the scientific world on the occasion of Tesla's seventy-fifth birthday, in 1931. He asked each one individually to pay homage. With the co-sponsorship of Dr. Bernard A. Behrend and Professor Charles F. Scott, more than fifty members of the American Institute of Electrical Engineers, including fourteen past presidents of that organization, sent their congratulations on the event.[1]

Tesla was surprised to find articles appearing in periodicals over this country and the world, commemorating his birthday anniversary. His portrait appeared on the cover of *Time* with an accompanying article which stated that after some difficulty the reporters had found Tesla at the Hotel Governor Clinton. His days at the St. Regis had been over for some time—in part because the maids had complained about the four pigeons which he appeared to be studying and which had perched on his roll-top desk. The difficulties with the hotel were apparently in part financial, too, since there is a record that Tesla was involved in a suit brought by the St. Regis Hotel Company for a judgment of $3,299 for non-payment of rent.[2]

The *Time* reporters found Tesla more of a recluse than ever. Indeed, when newspeople called and he didn't care to see them, he had a habit of locking himself in the bathroom and turning the water on loudly. Even though he was growing old, he scorned sleep and still went to bed at 5:30 a.m. and arose at

10:30 a.m., and he made it clear that he did not sleep all that time, but that much of it he rolled around thinking about his scientific problems.

His annual interview was usually a rehash of old ideas and accomplishments, but on this, his seventy-fifth anniversary, he put forth some startling new ideas. The reporter stated that, "For forty years he [Tesla] had been arguing that the whole earth was a powerhouse."[3] The report went on to record further advancements which Tesla proposed in his world power system—that for electrical power the generators might be waterfalls, coal mines, or any other means anywhere. The only possible drawbacks to its practicality would be the vast expense of installation and the fact that every powerhouse on earth would have to generate the same kind of current. And further, anyone could tap the current so that there could be no financial control.

And then Tesla went on to reveal his most recent dream of an entirely new source of power. He maintained that he was working on two things based on pure mathematics which Professor Albert Einstein had also attempted to explain, but Tesla boasted that his explanations were not so involved as Einstein's, albeit they concerned a totally new source of power. He said:

> The conception, the idea when it first burst upon me was a tremendous shock. I can only say at this time that it will come from an entirely new and unsuspected source and will be for all practical purposes constant day and night. The apparatus for capturing the energy and transforming it will partake of both mechanical and electrical features. At first the cost may be found to be too high but this obstacle will be overcome. The installation will be indestructible and will continue to function for any length of time without additional expenditures. It has nothing to do with atomic energy. There is no such energy in the sense usually meant.[4]

He told the reporters that he had already conceived a means that would make it possible to transmit energy in large amounts, thousands of horsepower from one planet to another absolutely regardless of distance, and then he continued:

> I think nothing can be more important than interplanetary communication. It will certainly come one day and the certitude that there are other human beings in the universe working, struggling and suffering like ourselves will produce a magic effect on mankind, the foundation of a universal brotherhood that will last as long as humanity itself.[5]

Tesla told the reporters:

> I have been leading a secluded life, one of continuous, concentrated thought and deep meditation. Naturally enough I have accumulated a great number of ideas. The question is whether my physical powers will be adequate to working them out and giving them to the world.[6]

The newsmen noted his paleness, but commented that he appeared to be healthy, as strong and alert as ever, but thin to ghostliness. And they wished they might have seen him as he used to be seen in his Colorado laboratory a generation before; "strolling or sitting like a calm Mephistopheles amid blazing, thundering cascades of sparks 30 feet long, Tesla currents alternating at such prodigious frequency that they would not harm a kitten."[7] They noted that his hair had become slate gray, overhanging eyebrows almost black—and the blueness of his eyes, and they commented, "Only their sparkle and the shrillness of his voice indicate psychic tension."[8]

Because of Kenneth Swezey's prodigious efforts, other articles appeared and letters began to pour in from all over the world. Among them was one which might have torn open old wounds. It was from William H. Bragg, who had been awarded the enigmatic Nobel Prize in 1915. Writing from the Royal Institution in London, Dr. Bragg recalled the wonders of Tesla's experiments of forty years before and said, ". . . I shall never forget the effect of your experiments which came first to dazzle and amaze us with their beauty and interest."[9]

The great names of the world of atomic physics were signed to some of the letters. Among them, Robert Millikan wrote:

> Dear Dr. Tesla,
> When I was a young man of twenty-five as a student in Columbia University I attended a down-town public lecture in New York at which you made one of the first demonstrations of your Tesla coil and its capabilities. Since then I have done no small fraction of my research work with the aid of the principles I learned that night so that it is not merely my congratulations that I am sending you but with them also my gratitude and my respect in overflowing measure.[10]

Another notable scientist from the same field, Arthur Compton, paid his respects, saying:

> . . . To men like yourself who have learned first hand the secrets of nature and who have shown us how her laws may be applied by solving our everyday problems, we of the younger generation owe a debt that cannot be paid . . . a man who taught us to use alternating currents more effectively for driving electric motors is by that one achievement entitled to the enduring gratitude of mankind.[11]

Active researchers in the field of radio technology were quick to acknowledge their debt to Tesla and to send their greetings. Louis Winslow Austin of the Research Council of the International Radio Union in Washington, D.C., said:

> I am glad to express my feeling of the great debt that radio art owes to your genius. The solution to many of the problems of the art which you presented at so early a date—1893—undoubtedly guided many of the later workers and laid a foundation for the large part of its present development.[12]

Louis Cohen, Consulting Engineer in Washington, D.C., paid his tribute to the pioneer work of Tesla in the field of radio, as did Ernest F. W. Alexanderson, from General Electric in Schenectady, who wrote:

> In almost every step of progress in electrical power engineering, as well as in radio, we can trace the spark of thought back to Nikola Tesla. There are few indeed who in their lifetime see the realization of such far flung imagination.[13]

And Lee De Forest, considered the father of radio, added a warm message from the American Television Laboratory, Limited, in Hollywood, commenting:

> My dear Mr. Tesla;
> On the occasion of your 75th birthday I wish to express my heart felt appreciation of my deep personal obligation for you as scientist and inventor.
>
> For no one so excited my youthful imagination, stimulated my inventive ambition or served as an outstanding example of brilliant achievement in the field I was eager to enter, as did yourself.
>
> Your simple statement on one occasion, that you "knew you could succeed," renewed my courage and gave me faith in myself at a time when I was sorely tried.
>
> I can freely state that one of the greatest satisfactions of my life has been the realization that your faith in me was not misplaced, and that I was finally enabled to achieve an invention which could be placed in a niche alongside of your great gift to humanity, the polyphase motor. Not only for the physical achievement of your researches on high frequencies which laid the basic foundations of the great industry of radio transmission in which I have labored, but for the incessant inspiration of your early writings and your example, do I owe you an especial debt of gratitude.

> That your remaining years may be many, and filled with the richness of a realization that your life has achieved greatly for humanity, is the sincere wish of an early disciple.[14]

The curator of the Radio Corporation of America Museum added his tribute and recalled the Colorado experiments and the work at Wardenclyffe, prophesying that soon the examples of the pioneer work of Tesla should be displayed in a prominent position in the Smithsonian Institute.

In a heavy script, Arthur Korn sent his message from Germany, paying Tesla honor as the pioneer instrumental in the development of modern photo-telegraphy.

A former assistant at Columbia University, Gano Dunn, testified to the indelible impression which Tesla's high frequency lectures had made upon him and expressed his gratitude for the inspiration which had influenced his life.

S. M. Kentner, vice president of Westinghouse Electric and Manufacturing Company, praised the genius of the polyphase system of power generation—transmission and utilization—first proposed by Tesla and developed by Westinghouse, a development which had started the electrical industry on its rapid march to its mammoth size, and he added that there were few pioneers in the electrical industry who could rank with Tesla.

D. McFarland Moore, of East Orange, New Jersey, recalled sitting beside Tesla on a platform of Columbia College on May 20, 1891, and a Sunday afternoon stroll up Fifth Avenue, three years later, when Tesla had stopped and said thoughtfully to him, "After we have signaled from any point of the earth, the next step will be to other planets." Moore commented on the suitability of the Tesla name cast into metal on the first water turbine alternating current generators at Niagara.

Charles F. Scott, of Yale Station, New Haven, Connecticut, recalled a Tesla lecture and said, "There he came, marching down the aisle with head and shoulders erect and with a twinkle in his eye. It was a great moment for me."[15]

Although there were names more famous than that of Tesla's old associate and defender, Dr. B. A. Behrend, no tribute could have been more sincere than his. He wrote:

> Among those who are happy to express their good wishes on your impending birthday, I am one of the humblest but perhaps the last living engineer whose good fortune it has been to design literally millions of horse-power of motors and generators of the "Tesla system." And though one should think that such monuments to your great inventive genius might be sufficient to fill the world with the fame of the name you bear it must be recorded that such would be the case had it not been for the world's usual ingratitude towards its benefactors.
>
> To those of us who have lived through the anxious and fascinating period of development of alternating current power transmission there is not a scintilla of doubt that the name of Tesla is as great here as the name of Faraday is in the discovery of the phenomena underlying all electrical work.[16]

Behrend, who could never forget the injustices which Tesla had suffered at the hands of Edison and others, added:

> Comparisons are odious and at the moment, while we are grateful that you have lived to see your great ideas realized far beyond your own dreams, it is perhaps bad taste to point out that the popular inventor of the incandescent lamp has had to look to your discoveries and inventions, and to our generating equipment designed with your ideas, for the generation of the electric energy without which his inventions would have remained lifeless.
>
> As there is so much popular misunderstanding on this subject I think these lines may help to adjudicate where fame should be and where praise should fall.[17]

Kenneth M. Swezey, who had sparked this anniversary recognition of international proportions, paid his own tribute to his friend in a news article. He spoke of the genius who might be a wizard and an ideal to some, a fanatical prophet to many. He praised Tesla for the startling impetus he had given to the work of J. J. Thomson, Roentgen, Millikan, and others and recalled the days when scientists had sought out the germs of the Tesla ideas and applied them to modern science. Swezey recognized the far projected vision of Tesla which had caused his being branded "mad man" by many, and an enigma to many more, but he asserted that the genius had "talked, written, and created as a poet." "Standing alone," said Swezey, "he plunged into the unknown. He was an arch conspirator against the established order of things."[18]

Hugo Gernsback, magazine editor, had long considered Tesla the world's greatest inventor—the greatest of all times—and placed him ahead of Archimedes, Faraday, or Edison. Gernsback had said:

> If you mean the man who really invented, in other words, *originated* and discovered—not merely *improved* what had already been invented by others, then without a shade of doubt, Nikola Tesla is the world's greatest inventor, not only at present but in all history. . . . His basic as well as revolutionary discoveries, for sheer audacity, have no equal in the annals of the intellectual world.[19]

When Swezey came proudly to call on Tesla and to present him with the shower of tributes, he found Tesla surprised, but not overly effusive. Tesla merely stated that he didn't care for compliments from people who had opposed him all his life, but secretly he was gratified, nonetheless, for when Swezey tried to borrow the letters later, for a short period, Tesla was reluctant to part with them and it was only after much urging that he released them for a time.[20] The year following the anniversary saw the letters published in Belgrade (Beograd) in a volume dedicated to Tesla.

Looking back over these tributes, Tesla may have recalled past honors and achievements, such as his lectures before the Royal Institute in London in 1892, the Society of Engineers in

Paris, and the Physical Society of Paris, as well as the Institute of Electrical Engineers in London (all that same year); his lecture before the Franklin Institute in Philadelphia in 1893, and the one in St. Louis for the National Electric Light Association, when the public had set up a clamor for Tesla; and he may have remembered his noted demonstration of 1888 before the American Institute of Electrical Engineers.

The acclaim of the world had been his at the Columbian Exposition, and with the development of the Niagara Falls power plant Tesla had been the man of the hour at the honorary banquet at the Ellicott Club. He had been awarded such past honors as a diploma, in 1908, from the scientific men of Austria for discoveries of the principles of transmission of power and of wireless telegraphy. In 1916, the American Institute of Electrical Engineers had presented him with the Edison Medal, and in 1926 he had received the Grand Cross of the Serbian Order of St. Sava from his native country.

If Tesla could have had the second sight of the Magyars of his birthplace, he could have forseen still more honors to come, such as in 1934 when the city of Philadelphia would award him the John Scott medal for his polyphase power system. And on his eightieth birthday, the opening of the Tesla Institute in Belgrade. Here, in the university's large auditorium, would gather seventy delegates, representing fifteen countries, convened to pay honor to Tesla. There would be papers on scientific subjects followed by the conferring of the Order of the White Eagle on Tesla by King Peter's representative, Regent Prince Paul. Tesla would cable his thanks with "Long Live Yugoslavia and its Royal House."[21] Not the least of the accompanying ceremonies was to be the visit of the famous scientists to the village of Smiljan in the Lika mountains where the villagers would share in paying honor to their native son at a gathering in the village church, a distinguished pilgrimage to his birthplace. And his native country would pay further honor with a postage stamp bearing his portrait.

And it would have been gratifying if Tesla could have foreseen the honorarium of $7,000 per annum voted to him by the Tesla Institute on the seventy-fifth anniversary. This could by no means be sufficient for one whose work required such expensive equipment that a single piece might well cost far more than the annual income. This fund, however, was to insure his essential comforts in his accustomed haunts in New York. His physical needs were growing less, but his hunger for a laboratory of unlimited scope was never to diminish.

A scroll of honor from the National Institute of Immigrant Welfare was to come to Tesla in 1938, as a foreign-born citizen whose influence was national and international in scope, constructive in character and purposeful in objective—an honor to be shared with Martinelli and Felix Frankfurter.[22] Had Tesla possessed the gift of second sight, he might have smiled at the cumulative eulogies which would appear at the time of his death. Obituary writers, declaring him a Jules Verne character, were to revive Tesla's insistence on his communication with Mars, his ability to pluck heat units out of the air, and his use of the whole earth as an electrical resonator.[23]

It would have been gratifying to him to see the newsmen who had so often ridiculed him, come to admit that his thinking had been undergirded by a solid scientific foundation, that he was no tinkerer, and that if the abused word *genius* was ever applicable to any man it was to him.[24]

And how pleased Tesla's old friend, Robert Underwood Johnson, would have been to have peered into the future with him to see, on June 29, 1952, the unveiling of a bronze replica of a bust of Tesla at the Technical Museum at Vienna, Austria. The original, sculptured by the noted Ivan Mestrovic, after Tesla's death would remain on exhibition at the Yugoslav Academy of Art and Science in Zagreb.

Again, on the 100th anniversary of Tesla's birth, was to come another flurry of remembrance when scientists and engineers again would gather to pay homage to his memory. A year of commemorative programs was to highlight 1956, stemming from the activities of the Nikola Tesla Museum in Belgrade, under the auspices of the National Yugoslav Committee, the Society for the Promotion of Science and Technology. Among the distinguished guests who were to gather in Belgrade were Neils Bohr of Denmark, Arthur P. M. Fleming of England, Frederik Dahlgren of Sweden, and Carl Chambers and Richard

Sogg of the United States, representing the American Institute of Electrical Engineers and bringing a citation to the Tesla Committee from that organization.

In Munich, the Institute Electrotechnical Committee was to make a formal agreement to adopt the name "Tesla" for the unit of magnetic flux density in the MKS or Giorgi system, thus placing Tesla alongside Ohm and Ampere.[25]

The festivities and honors were to continue with the dedication to Tesla of the fall meeting of the American Institute of Electrical Engineers in Chicago, October 1 – 5, 1956. Sogg was to report on the lectures in Belgrade with the accompanying visits to technical and industrial plants and to describe such memorable festivities as gypsy dances, a theatrical evening, and a sail on the Danube, with the announcement of a crowning achievement, the publication, in English, of the volume of Nikola Tesla's lectures, patents and articles.

Chicago would appear the logical place for the opening of such a celebration for it was here that, in 1893, the Tesla polyphase system of power transmission was demonstrated at the World's Fair, and it was in the same year that the Chicago section of the American Institute of Electrical Engineers, the oldest chapter in the United States, had been organized. Here, too, Tesla had demonstrated his first radio-guided weapon, in 1899, en route to Colorado Springs. He had returned in 1917 while engaged in work on mechanical transformers.

In connection with the 100th anniversary celebration in Chicago a bust of Tesla was to be on view, one completed only a short time previously by Stephen Segvich. Through the courtesy of one of the publications which had damned Tesla in 1900 (*Popular Science*) was to appear a large picture-narrative of his life. Also was displayed a model of Wardenclyffe Tower and three large photo enlargements of Tesla at the age of seventy-seven, a display to be prepared by the AIEE. Accompanying the photographs were to be models of his guided weapon and replicas of the cascading electrical discharges of his Colorado Springs experimental station. The Mayor of Chicago would dedicate October 1 as Nikola Tesla Day.

Dr. Hibben of Chicago would be the speaker to forecast that "the lengthening shadow of Tesla's passing will yet touch and inspire us to a greater future accomplishment than we have ever known."[26]

And still more honors were to come from Yugoslavia, not from the royal family, but now from its ruler Josip Broz (Tito). The Secretary of State of the United States Department of State was to write Tito concerning the proposed commemorative festivities and a reply would read:

> July 16, 1956
> Dear Mr. Secretary:
> I have the honor to transmit to you herewith the following message of the President of Yugoslavia, Josip Broz Tito.
>
> "Dear Mr. Dulles:
> I have received your personal message through our embassy at Belgrade on the occasion of the celebration of the 100th anniversary of birth of the renowned scientist Nikola Tesla. Thanking you for your amiable words on the occasion of this event which is of particular significant [sic] for our two countries, I would especially like to stress my accordance with your statement that scientists of Tesla's genius are the symbol of the universality of science and human endeavor for progress in peace.
> With kind regards, Josip Broz Tito"
>
> Please accept, dear Mr. Secretary, the assurances of my highest consideration. Franc Primozic, Charge d'Affaires.[27]

If Tesla could have foreseen these cumulative honors gathered throughout the many years, he would have smiled, perhaps, but he would have known that every such honor was well deserved, and that in return he could have given much more to the world, if only they had allowed it.

DIPLOMAS AND OTHER AWARDS

(from *Tribute to Nikola Tesla, Letters, Articles and Documents,* Beograd, Yugoslavia: Nikola Tesla Museum, 1961, pp. xix-xx.)

Certificate of the Elliott Cresson Gold Medal Award (The Franklin Institute in Philadelphia, 1893)

Diploma of honorary Doctor of Collegium Columbiae (1894) Diploma of honorary M.Sc. of Yale University, Connecticut (1894) Certificate of election as Fellow of the American Association for the Advancement of Science (1895)

Certificate of election as honorary Fellow of the American Electro-Therapeutic Association (1903)

Certificate of election as Active Member of the New York Academy of Sciences (1907)

Diploma of honorary Doctor of Technical Sciences of the High Technical School in Vienna (1908)

Certificate of the Edison Gold Medal Award (The American Institute of Electrical Engineers) (1916) (sic) [1917]

Diploma of election as Fellow of the American Institute of Electrical Engineers (1917)

Diploma of honorary Doctor of Technical Science of the University of Beograd (1926)

Diploma of honorary Doctor of Technical Sciences of the University of Zagreb (1926)

Certificate of the John Scott Medal Award (The City of Philadelphia, 1934)

Diploma of honorary Doctor of Technical Sciences of the High Technical School in Prague (1936)

Diploma of honorary Doctor at the University of Poitiers (1937) Diploma of honorary Doctor of Technical Sciences of the High Technical School in Graz (1937)

Diploma of election as Regular Member of the Serbian Academy of Sciences in Beograd (1937)

Diploma of honorary Doctor of Technical Sciences of the High Technical School in Brno (1937)

Diploma of honorary Doctor of the University of Paris (1937) Diploma of honorary Doctor of the Polytechnical School in Bucharest (1937)

Diploma of honorary Doctor of the University of Grenoble (1938) Diploma of honorary Doctor of Physical Sciences of the University of Sofia (1939)

Scroll awarded by the Technical Museum in Vienna (1952) Scroll awarded by the *Popular Mechanics* Magazine on election among ranks of the 50 most famous Americans in the field of science and invention during the last 50 years (1952)

Scroll awarded by the American Institute of Electrical Engineers (1956)

Tribute to Nikola Tesla Centenary (City of Philadelphia, 1956)

Resolution by the Board of Menagers (sic) of the Franklin Institute (1957)

Document of the International Electrotechnical Commission (IEC) on the adoption of the name "Tesla" for the unit of magnet induction (1956)

Photograph of the front of the Head Building of the "Electricite de Strasbourg"

Verbal Tributes to Tesla

> Genius is its own passport, and has always been ready to change habitats until the natural one is found. Thus it is perchance that while some of our artists are impelled to set up their easels in Paris or Rome, many Europeans of mark in the fields of science and research are no less apt to adopt our nationality, of free choice. They are indeed Americans born in exile.
>
> Thomas Commerford Martin,
> in *Century Magazine* February, 1894, p. 582.

> He beat new paths to mountains, dared to turn from his father's profession, defied his conventional professor, dared to defy the electrical wizards of his day . . .
>
> John Patterson

> It was fortuitous that the discoveries of this great genius, which revolutionized generation and transmission of electrical power, should be available just when the need as shown by the Niagara Falls project was most pressing.

In the growing field of electrical application Tesla's ingenuity and incessant energy found full scope and his activities touched every phase of electrical engineering and overshadowed those of any other electrical engineer of his time.

Arthur P. M. Fleming,
from Lecture read at the Tesla Congress
in Beograd, 1956.

The evolution of electric power from the discovery of Faraday in 1831 to the initial great installation of the Tesla Polyphase system in 1896 – undoubtedly the most tremendous event in all engineering history.

Charles F. Scott,
in *Electrical Engineering,* August 1943, p. 351.

On this anniversary of your birth, the Niagara Falls Power Company offers congratulations to the man who some thirty years back wrote after his name in its visitors' book, "Stop in New York but heart is at Niagara," and who so fundamentally contributed to the world's pioneer power development.

. . . It is fitting that your name continue as it now is, inscribed on the bronze nameplate of the first 5000 h.p. alternator used at Niagara.

Paul A. Schoellkoff,
President of Niagara Falls Co., Letter, July 10, 1931

Tesla, by virtue of the amazing discoveries and inventions which he showered on the world, becomes one of the most resplendent flashes that has ever brightened the scroll of human advancement.

John J. O'Neill,
Prodigal Genius

. . . Nikola Tesla happened to be one of the world's greatest geniuses. Father of the alternating current power system, a multiphase current transmission of high voltage, he ushered in the modern electrical power era of today. Through his discoveries, Boulder Dam was to be built and paid for and the Colorado River finally controlled.

Frank Waters,
The Colorado, p. 333

Tesla was born in the covered wagon age, lived to see the world of electricity, but most of all he planned it and engineered it.

Gardner Soule,
in *Popular Science,* July 1956, pp. 81-85.

. . . Tesla would see multitudes of possibilities for investigation . . . and his genius could not be confined any more than a mighty stream.

W. W. Wilhelm,
from *Nasa Posta*

The progress of the world is so heavily dependent upon the development of the Teslas that the slightest progress in improving their education would be of inestimable value.

Carl C. Chambers,
from Lecture read at Tesla Congress, 1956

Tesla once confided to me, "I will never have any money unless I get it in amounts so large that I cannot get rid of it except by throwing it out the window."

Kenneth M. Swezey,
from "Nikola Tesla" in *Science,* May 1958, p. 1147.

. . . the apparent idealist is he who often points the way for others, and therein is the secret of the noble aspirations and greatness of Tesla.

E. J. Holmgren,
in *Nature,* London, 1956, p. 1426.

He joined the insight of a poet with the patience of a seeker of exact truth, quickened by imagination—a scientist, an inventor and a seer all in one and one of the most remarkable men of this day or any other time.

John Ford,
in *New York Times,* September 30, 1894, p. 20, col. 1.

Notes

1. Kenneth M. Swezey, "Nikola Tesla, Pathfinder of an Electrical Age," in *Electrical Engineering,* September 1956, 4.
2. *New York Times,* May 28, 1924.
3. "Tesla at 75," *Time,* July 20, 1931, 27-30.
4. lbid.
5. lbid.
6. lbid.
7. lbid.
8. lbid.
9. Kenneth Swezey, ed., *Nasa Posta,* Volume dedie au jubilee de Nicolas Tesla, (Beograd, Yugoslavia: 1932).
10. lbid.
11. lbid.
12. lbid.
13. lbid.
14. lbid.
15. lbid.
16. lbid.
17. lbid.
18. Kenneth M. Swezey, "Nikola Tesla Envisioned Present Electrical Era 40 Years Ago," *New York Herald Tribune,* July 19, 1931, 1-2.
19. Hugo Gernsback, "Nikola Tesla and His Achievements," *Electrical Experimenter,* January 1919, 614-615.
20. Interview with Kenneth M. Swezey.
21. *New York Times,* May 29, 1936.
22. Kenneth M. Swezey, "Nikola Tesla," *Science,* May 1956, 1157.
23. *New York Times,* January 9, 1943.
24. Ibid.
25. Kenneth M. Swezey, "Pathfinder," 5.
26. "Nikola Tesla Centennial Observance," presented by Leland Anderson of the Tesla Society.
27. From correspondence with E. Taylor Parks, officer in charge of Research Guidance and Review, Historical Office, Department of State, Washington, D.C.

TESLA, AT 78, BARES NEW 'DEATH-BEAM'

Invention Powerful Enough to Destroy 10,000 Planes 250 Miles Away, He Asserts.

DEFENSIVE WEAPON ONLY

Scientist, in Interview, Tells of Apparatus That He Says Will Kill Without Trace.

New York Times, July 11, 1934.

Tesla's Death Ray & the Projection of His Ideas Into the Future

"The past is dead, the present is an enigma, and the future is a riddle." How much had Tesla unraveled of that enigma in the thirties, and how many riddles of the future had he solved in his own mind? In his later years, he and his work began to slip further and further into obscurity. It was said of him:

> A man of flashing insight and enormous brilliance, Tesla was largely indifferent to the development of his ideas. This he left to others while he followed the lure of new challenges. In later years his projects became more grandiose, his ways more mysterious, his pronouncements more Olympian. And working alone, as he did, he formed none of the institutional ties that helped to perpetuate a record of accomplishment.[1]

Tesla did not see eye to eye with Einstein and his modern theory of relativity. However, Tesla accepted the existence of the electron, but not as associated with any part of an atom. "Atomic power is an illusion," he declared and insisted that he had many times smashed billions of atoms but that little energy had emitted from his process.

Friends who knew him in his later years felt that he became less dogmatic concerning the theories of others. He confided to them that he was contemplating an apparatus for testing the theory of atomic structure, but that his device would release atomic energy more effectively than the methods then currently in use. His ideas had been so often pirated in the past that he was wary of any revelation and he had no laboratory equal to carrying through his new ideas. If Tesla and Einstein could have combined their monumental powers it would have been to the advantage of the world, but both were eccentric—and Tesla was slipping from the public acceptance.

His old friend Mark Twain had coined the expression, "Every one talks about the weather, but no one does anything about it." But to Tesla, weather was a serious science and he was certain that man could do something about it to his own advantage. As early as 1914, he predicted electrical control of atmospheric moisture and the possibility of drawing unlimited quantities of water from the oceans. He foresaw the safeguarding of forests against fires, the destruction of microbes, insects, and rodents—all by electrical means.[2]

In an unpublished article he ventured to predict what would happen if Man could cause nature to operate by the force of his will—old worlds would disappear and new ones would spring into being. Tesla maintained that man

> could alter the size of this planet, control its seasons—guide it along any path he might choose through the depths of the Universe—he could cause planets to collide, and produce his suns and stars, his heat and lights. He could originate and develop life in all its infinite forms.
>
> To create and to annihilate material substance, cause it to aggregate in forms according to his desires would be the supreme manifestation of the power of Man's Mind—his most complete triumph over the physical world and his crowning achievement which would place him beside his Creator, make him fulfill his ultimate destiny.[3]

Did he foresee, three decades later, a research project known as "Sky Fire" would be engaged in studying the electrical characteristics in the Rocky Mountains and the evaluation of the effects of cloud seeding on lightning occurrence? Could he have caught a faint glimmer of a military investigation when thousands of volts of electricity would be used to dissipate a shallow artificial fog layer?[4]

And what of the death ray? As early as May 19, 1924, the British scientist, Grindell H. Matthews, was quoted from a Paris dispatch as the discoverer of a diabolic ray capable of destroying airplanes and it was announced that he would be coming to Paris on the following day to join the French scientists.[5] On May 21st, Paris again sent out word that Grindell Matthews had knowledge which would make it possible to put a whole army out of action and that rays would be used to direct an electric current.[6] However, scientists deplored the exaggerated accounts of his experiments, saying that a four-mile radius was possible with a maximum of seven to eight miles, thus rendering the destruction of an entire fleet impossible. Matthews insisted that the Germans already had such a ray.

The following day, Dr. W. L. Severinghaus, a physicist at Columbia, said that Matthews' claims were "fishy" and he tried to still the fears of an agitated public.[7] By May 25th, Dr. T. F. Wall, lecturer in electrical research, had applied for a patent, claiming the death ray discovery, insisting that he could transmit electrical energy without wires, stop airplanes and cars, and that the invention could be of use in surgical and medical fields.[8]

Everyone was afraid of being sold out to the enemy.

On May 25th, a Herr Wolle announced that the Germans had three inventions of death rays which he termed the curtain of death.

Three days following, the British were very concerned whether Matthews might sell out to the French, an act which might be construed as being both indiscreet and disloyal. Matthews had been offered one thousand pounds if he could put a small motor out of business with his invention. He left for Paris with no deal concluded and the tests went on everywhere, but none was conclusive. It was stated that Matthews had received an offer from the United States Navy, but that he was using a foreign bid to boost the price. It was even suggested that behind it all lay Trotsky's recent warlike utterance claiming an electromagnetic invention by a Russian engineer, Grammachikoff, an invention which had been tested on a large scale. It was believed that the Russians were planning to construct antiaircraft based on these plans.[9]

For years, common gossip had attributed a mysterious death ray to Tesla's inventive genius. Not to be overlooked in the headlines of the day, Little London claimed proud relationship to Nikola Tesla and his Colorado Springs discoveries which they had somehow overlooked at the time, but now claimed "they knew him when—." On May 30, 1924, a Colorado Springs paper carried on its first page a news article inspired by a New York report of the day before. The story told of the invention of an invisible ray, developed by Tesla, which was capable of stopping airplanes in midflight, an invention which had come about through improvements on Tesla's Colorado Springs discoveries. It was boasted that the ray had already been used to cause French airplanes to descend over Bavaria, and went on to state that the ray had been offered to the United States government through an English engineer, J. H. Hammil, who represented a German scientist who had perfected the ray.[10] It was mentioned that Tesla had offered to turn his death ray plans over to Geneva.

On his seventy-eighth birthday, in 1934, another flurry of interest in Tesla's death ray hit the country. Again, an Associated Press dispatch picked up the reporter's story. Here, Tesla was quoted as saying that his ray was based on an entirely new principle of physics. He upped the figures, claiming that his beam could destroy 10,000 planes at a distance of two-hundred and fifty miles. Although the beam was only one one-hundred-millionth centimeter, the inventor admitted that a two-million dollar plant would be required for its construction and that the time necessary for its completion would be three months. Tesla stated that twelve such plants, located at strategic points, would constitute a veritable Chinese wall around the country, and could defend the United States against all foes. His beam, he claimed, could melt any engine, whether driven by diesel or gasoline or oil, and there was no possible defense against it. Tesla insisted that he could go to work at once, but that he would "tolerate no interference from experts."[11]

The ray was described as the most important of all Tesla's inventions so far; it was said that it could send concentrated beams of particles through free air, could cause armies of millions to drop dead in their tracks. It was stated that the beam would operate silently but effectively, as far as a telescope could see an object, or on the ground as the curvature of the earth would permit, and would leave no trace of what had caused destruction. The Tesla ray involved four new inventions:

> one, comprising an apparatus for producing rays and other manifestations of energy in free air, eliminating the high vacuum necessary for producing such rays and beams; the second, a method for producing great electrical force; and the third, a plan for amplifying this process in the second invention;
>
> and the fourth a new method of producing a tremendous repelling force.[12]

The voltages to be employed were fifty million volts, catapulted for destruction. The entire process was labeled, by Tesla, his "teleforce."

What happened to the death ray idea? Who knows? Did Geneva pass it off as the babblings of senility? Did the United States government, secure in its natural barriers of oceans and the certainty of the eternal friendship of neighbors with the unguarded borders of north and south, of the eternity of an inviolable Monroe Doctrine, plus the deterrent of air power, and an eye toward the economy based on spiraling expenditure, discard the idea? Or was it pigeonholed in the classified files? Did the inventor hold the blueprint only in his mind or did he sketch it on a tablecloth—or was it one of the scraps of paper squirreled away in his trunks?

In 1938, in a paper read *in absentia,* on the occasion of his being awarded an honorary citation, he again described his death ray beam as being so powerful that it could be flashed through interstellar space to any distance and substantiated his previous claim that this powerful ray could produce an incandescent spot in the dark regions of the new moon which would glow so brightly as to be seen without a telescope.[13]

Two years later he was still making claims for his death ray to a disinterested public.

Could this have been the idea from which came the birth of the maser from Dr. Charles Hard Townes, then a young scientist with Columbia University (now with the Institute of Defense Analysis in Washington, D.C.)? While waiting for a restaurant to open one morning he sat on a park bench pondering the problem of how to produce microwaves shorter than any other previously generated. Before the restaurant had opened, Dr. Townes had the germ of an idea which, after four years of experiments, led to the device which produced the maser (Microwave Amplification by Stimulated Emission of Radiation).[14]

From the development of the maser sprang the idea for the laser ray (Light Amplification by Stimulated Emission of Radiation) developed by Dr. Theodore H. Maiman, then of Hughes Aircraft Company.

In 1962, a team of Massachusetts Institute of Technology engineers, headed by Professor Louis Smullin and Raytheon Engineer Glen Hardway, by means of the new light device, the laser, bounced a beam off the moon. When the laser emitted its light, the telescope directed a beam twelve inches in diameter, at the dark half of the moon. Two and a half seconds later the returning light, less than a ten billion billionth of the light sent, streaked into the top of the telescope. Each flash had been focused on the target mountains near the crater Albategnius, the Copernicus crater. An ordinary beam would have spread thousands of miles and then been lost.[15]

The light produced by the laser is a fleeting red flash, an extraordinary kind of light never found in nature. It is "coherent" light and does not spread out, as does natural light. The rays stay parallel over vast distances. The burst of the red laser light can blast a pinpoint hole through a razor blade in a thousandth of a second and can heat steel to 10,000 degrees Fahrenheit. This is the simplest kind of light, containing only a single color, with waves of the same length and moving in phase with each other. It can travel thousands of miles without spreading out significantly. The heart of a laser is a four-inch ruby rod around which coils an electric flash lamp. (The laser ray is no longer restricted to the ruby ray.) A bright pulse of ordinary white light from the spiral bulb causes the ruby rod to emit an intense burst of ruby red light. The ultimate source is chromium, the element which gives ruby its color. The chromium atoms absorb ultra-violet, or green and yellow light, and through juggling electrons, convert it to red light. The atoms are forced into a single powerful beam into the ruby rod. Both ends of the rod are polished to act like mirrors. Aimed at either end, the light is reflected back and forth in the crystal and the atoms are stimulated to give off more rays, thus producing the action.

In a few billionths of a second, such a powerful beam is built up that it bursts out of the right end of the rod, which has been left more transparent than the other.[16]

Could this magic ray have anything to do with Tesla's death beam which was described as being infinitesimally small? Did the idea of aiming the laser at the moon have its origin with Tesla, who predicted that he could make a spot glow on the dark side of the moon?[17] Was Tesla overestimating the ability of his ray whose effect he said could be seen without the aid of a telescope? Is there a chance he had made the right prediction, which may yet be verified in the future?

Could Tesla have known the secret of the ruby light? Or was this principle, which he said was entirely unknown in the field of physics of his time, a completely different idea, yet to be discovered? Or was the reporter right who declared Tesla's predictions the babblings of senility?

Tesla would have been pleased to see scientists investigating this newest light device for medical research, metal working, radar, satellite tracking, and power transmission and communication.

In the field of medicine, high hope is held for the laser's efficacy in eye surgery for retinal detachment. To Tesla, the retina of the eye was all-important and the repose of many secrets. He would take great interest in the anticipated uses of the laser for cauterizing wounds, for delicate brain surgery, and, in the realm of biology, for controlling heredity and for the study of human tissues. Metal workers are already contemplating computer devices powered by lasers to eliminate many machine tools. Radar may be revolutionized through use of the newly discovered ray because it is a more powerful and more accurate source than any other yet found. It is proposed to apply it for satellite tracking since it would enable space ships to rendezvous in orbit "with hardly a bump."[18]

In the field of communications, the laser could transmit more information than all present radio channels. The ruby ray here, too, may open vast new realms. One writer maintains that just one laser beam could carry 25,000 simultaneous television programs, could make possible communications between submerged submarines, transmit power without wires, and send hundreds of times more messages than all undersea cables now in, existence.[19] Information could be stored in one crystal or in tiny laser units. Instead of electronic tubes and transistors, computers would be designed to use the laser for storing information.[20]

Since radio waves are useless under water, the laser would meet some difficulties. As a result, a new green light laser is being developed, since green is least absorbed by water. Then, too, the problem of cloud formation and rain would require that laser beams might have to be conducted through pipes, but there are no weather problems in space so that the long distance laser beams might prove ideal for interplanetary telephone calls.

Since the new light first appeared in 1960, some 400 firms and universities have launched research projects in laser, and an estimated twenty-million dollars was spent in 1961 for laser experimentation. Thirty-one uses are predicted for the next decade.[21]

Billions of times hotter than the sun's surface, the rays have punched holes in diamonds. Their force may make the electric light obsolete. The tiny speck of metal, indestructible, will never burn out. Could this have been connected with Tesla's prediction that he could light the ocean at night?

Government and military authorities are using laser experiments to extend their military power. Russia is not ignoring the latest developments in light rays. Of necessity, much of the progress is classified information.

So the small weapon is becoming a threat on both sides. It ranges in size from not much larger than a flashlight to the size of a cannon. It was reported that a manned anti-missile satellite is already under development by the Martin Company and that their goal is to produce a fifteen-ton vehicle containing a nuclear reactor and some other power source which can generate a continuous laser beam with a temperature of about one million degrees. A radar system in the satellite would detect any enemy as it rises above the atmosphere. Another radar would focus the laser beam on the enemy satellite and disintegrate it. One writer is pessimistic about the technical possibility of laser development and takes a dim view of such too-glowing prophecies which have stated "if laser techniques advance as rapidly as expected, lasers may replace almost all existing weapons from hand guns to air defense missiles and anti-missiles."[22]

Others have announced that secret military objects are underway, and many experts predict that laser bullets of light may usher in a kind of warfare so far envisaged only in science-fiction.[23]

Hughes Aircraft researchers are among those prophesying that the laser may be the solution for anti-missiles; a super-powered version of the laser's electronic lightning may burn holes in attacking missiles and turn them into flying scrap metal. It is said the ray could knock out trucks, tanks and foot soldiers, that it would not be necessary to destroy the entire missile since it would burn itself up on re-entry.[24]

What might have been Russia's picture today if Tesla had accepted an invitation from Vladimir Ilyich Lenin to work for them? Among old letters was found information that twice in succession, Lenin had made Tesla very tempting offers, but that he could not tear himself away from his laboratory at the time.[25]

Could he have planned something similar to the ruby light in the field of communications when, from his Wardenclyffe Tower, as early as 1900, he said he could send as many messages through the earth with his world system without interference? Had he ever conceived such an idea as using a device resembling a laser when he said he could supply heat or motive power anywhere on sea, land, or high in the air?[26] And what of his communication with planets?

Dr. C. H. Townes, who first constructed a maser, recommends that:

> . . . we use these new super light-masers to get in touch with intelligent races on other planets. Their fantastic

> power would make them our best bet for signaling over intergalactic distances.

And he dares to suggest that perhaps we should take a careful look around the sky for laser flashes right now. "Maybe somebody out there is already flashing our way," he says.[27]

Those who laughed at Tesla would not now have the temerity to take lightly the fantastic statements of a latter-day scientist such as Dr. Townes.

The man who said he could protect the United States with twelve stations emitting light beams which could burn to ashes, metal at two-hundred miles is now himself a burned-out genius whose ashes repose in a communist-held country. He is almost forgotten in his chosen land. And the popular periodicals of the day mention Buck Rogers and Flash Gordon, and credit their originators with cosmic imagination.

A little satellite named Sputnik turned the eyes of the world in a different direction. There are no holds barred and there is no limit to man's vision, today.

Long ago, in 1899, Nikola Tesla sat alone in the night hours in his experimental laboratory on Knob Hill in Colorado Springs and was certain he received communication from outer space. He was ridiculed by the world for his effrontery, but the lone scientist forgave the crass and unseeing.

Undismayed, he said at one time:

> The scientific man does not aim at immediate result. He does not expect that his advanced ideas will be readily taken up. His work is like that of the planter—for the future. His duty is to lay the foundation for those who are to come and point the way. He lives and labors and hopes with the poet [Goethe] who says
>
> > "Lo! these trees, but bare poles seeming
> > Yet will yield both fruit and shelter!"
> > (from Goethe's "Hope" translated by William Gibson)[28]

Notes

1. Louis Rowley, "Strange Genius," in *Power,* May 19, 1935, 67.

2. *Nikola Tesla, 1856 – 1945, Lectures, Patents and Articles,* (Beograd, Yugoslavia: Nikola Tesla Museum), A-182.

3. John J. O'Neill, *Prodigal Genius,* (New York, N.Y.: Ives, Washburn, Inc., 1944), 251-252.

4. *Colorado Springs Free Press.*

5. *New York Times,* May 19, 1924.

6. Ibid., May 21, 1924.

7. Ibid., May 22, 1924.

8. Ibid., May 25, 1924.

9. Ibid., May 28, 1924.

10. *Colorado Springs Gazette,* May 30, 1924.

11. *New York Times,* January 8, 1943.

12. Ibid., July 11, 1934.

13. O'Neill, *Prodigal Genius,* 242-243.

14. *Time,* January 2, 1961, 43.

15. *Life,* January 11, 1963, 53 ff.

16. Ibid.

17. O'Neill, *Prodigal Genius,* 242.

18. C. P. Gilmore, "The Invisible Ruby Ray," in *Popular Science,* September 1962, 88-92, 200.

19. Ibid.

20. *Science News Letter,* November 24, 1962, 333.

21. Harland Manchester, "Light of Hope or Terror," in *Reader's Digest,* February 1963, 97-100.

22. William E. Boggs, "Bullets of Light," in *New Republic,* March 16, 1963, 5.

23. Manchester, "Light of Hope or Terror."

24. Gilmore, "The Invisible Ruby Ray."

25. Microfilm letters, Tesla to Morgan, November 29, 1934, Library of Congress.

26. O'Neill, *Prodigal Genius.*

27. Gilmore, "The Invisible Ruby Ray."

28 Nikola Tesla Museum, *Lectures, Patents and Articles,* A-151-152.

Sava N. Kosanovic (Nikola Tesla's nephew and the ambassador to the U.S. from Yugoslavia), Nikola Tesla (age 86), and exiled 19-year-old King Peter II of Yugoslavia. Hotel New Yorker, July 15, 1942, six months prior to Tesla's death. *From Kenneth M. Swezey Papers, Archives Center, National Museum of American History, Smithsonian Institution (AC0047-0000017).*

Death of Tesla

Tesla's health was never quite the same after a taxi accident in 1937. He had refused all medical attention at the time and had tried to shrug it off as a minor circumstance. His heart bothered him occasionally as the years wore on and his more frequent fainting spells became an inconvenience and annoyance but nothing that he felt could not be lived with. Pneumonia struck a few vicious blows and left him weak. As time passed, his voice grew feeble and his words a little less coherent. He became more and more convinced that by sheer will power he would be able to control his physical well-being and that it would be possible to parry Death and to ward off the Dark Intruder as long as he wished.

His germ phobia grew more pronounced. He backed away from human contact, allowing very few people to come nearer than within a few feet of him. All his eccentricities became accentuated.

Christmas of 1942 came and the Great White Way of Broadway was even brighter because of the holiday lights. Radios blared *Silent Night* on the street corners. Neons flashed their red and green commercial greetings. The harassed shoppers were callously indifferent to the aged inventor who lived on the 33rd floor of the Hotel New Yorker, and who had changed the pace of their world.

George Scherff was one of the few friends with whom he could still communicate his dreams of new inventions. Scherff urged him again to take one idea and develop it into a paying proposition, but Tesla was obdurate. He felt the pressure of Time. There were still so many fields to explore and too many veils to be torn in the pittance of life left to him. Occasionally, a windfall came—but never enough. When he might have invested these bits in something profitable, he continued to pay off old obligations first. He was honorable but impractical.

The Christmas lights were still bright as Tesla turned his calendar to the year of 1943. On January 4th, Scherff came to

Tesla's office to assist him with an experiment. The weather was stormy and overcast. Barometric pressure had been falling and a few snow-flakes filled the air. New Yorkers turned their coat collars higher as the temperature fell. Sleet iced the streets. There is an account which says[1] that Tesla went to the window to watch the electrical flashes of an unseasonable thunderstorm[2] that competed with the neon flashes in the street below. "I have made better lightning than that,"[3] Tesla was said to have commented, as he watched the wintry sky. The inventor had been born in an electrical storm. He had devoted his life to the study of this mysterious power. In Colorado Springs, he had "tamed the wild cat" on an autumn night.

Suddenly he clutched at his heart as a sharp pain stabbed at his chest. Scherff begged to be allowed to call a doctor but Tesla refused. In a short time he rallied, rose from the couch where he had rested for a while, and insisted on going home alone. He even stopped on the way to order food for the pigeons.

The next day he allowed the maid to straighten his room but he gave her strict orders to keep visitors away and to see that he was not disturbed.

January 6th came and with Twelfth Night, the last hours of the Christmas season, the holiday lights dimmed over the world. The *Do Not Disturb* sign still hung on his door. On January 8th, Friday morning, the maid, becoming anxious about Tesla, ignored the sign, risked his displeasure and entered the room. Tesla had died quietly and alone during the night of January 7th. He had been dead for hours.

Immediately after the announcement of his death, his papers were impounded by the Custodian of Alien Property. According to Dr. H. Alan Fry, formerly chief of the Division of Foreign Languages, (OWI) (presently with the National Aeronautics and Space Administration), the papers were stored and are probably still kept in a warehouse in Brooklyn.[4]

Among those who gathered in the hotel room to take charge of affairs was Kenneth Swezey, who states that in Tesla's safe were his American citizenship papers, the letters from the 75th anniversary celebration, the Edison Medal, and other medals and diplomas. Of Tesla's entire estate only about $2,000 remained.

A team of engineers was called in to look over papers. They hastily weeded out those pertaining to their field. Personal papers were set aside to be housed later in the Tesla Museum in Beograd. Sava Kosanovich, Tesla's nephew, who later received the papers, estimated that it would take years to evaluate them. A few meager possessions were put into storage and there was no money to get them out.

Some biographers say that the operators from the Federal Bureau of Investigation opened the safe and removed the papers, "to examine them for a reported secret invention of possible use in the war,"[5] but those who were actually present agreed that it was the Custodian of Alien property, instead, who took charge.[6]

Tesla's thin gaunt body was taken to the Campbell's Funeral Parlors at Madison Avenue and 81st Street.

Hugo Gernsback, editor and friend, and long time admirer of Tesla, hurried to have a death mask made. Every facial contour and every fine wrinkle was faithfully reproduced. Gernsback had the mask copper-plated and placed in his office. There he planned to keep it as it acquired a soft green patina. On Tesla's one-hundredth anniversary in 1956, it would be a good plan to invite every associate and friend of the inventor to the unveiling of the mask.[7]

Kenneth Swezey was called in to help settle the estate. He remembers that the belongings were temporarily sealed, but later opened, and that they now remain open in the Tesla Museum in Beograd.

Because the splendid Serbian Cathedral was not completed at that time, the funeral services were arranged to be held in the Cathedral of St. John the Divine. This was done at the request of the Serbian Orthodox Church authorities. Because of the conflict between the Serbians and the Utashi[8] (an organization of Croatian Nationalists who desired separation from Yugoslavia), there was a tremendous emotional tension between the two groups. For this reason, Bishop William T. Manning gave his consent for funeral arrangements in the cathedral with the understanding that there were to be no political speeches made.

Telegrams poured in from everywhere with news of the death of this world citizen. Three Nobel Prize winners in physics paid tribute to Tesla as "one of the outstanding intellects of the world who paved the way for many of the important technological developments of modern times." This was the praise of Professor Robert A. Millikan, Arthur Compton, and James Franck.

Mrs. Franklin D. Roosevelt wrote in behalf of herself and the President of the United States:

> The President and I are deeply sorry to hear of the death of Mr. Nikola Tesla. We are grateful for his contribution to science and to this country.[9]
>
> Vice President Wallace took time to write a sincere expression of loss, saying:
>
> . . . in Nikola Tesla's death, the common man loses one of his best friends.[10]

The service was set for four o'clock on Tuesday, the twelfth of January. It was conducted by the Very Reverend Dushan Shoukletovich, who had known Tesla personally,[11] and who officiated in the name of the Serbian Orthodox Church of America. Assisting in the service was Father Edward West, Sacrist of the Cathedral.[12]

More than two thousand people filed into the aisles of the church, and among them came the great in science to pay their last tribute. Oscar Gavrilovitch, Yugoslav Consul in New York, headed the list of ushers who guided the throng to seats. Although Tesla had been a citizen of the United States for over a half-century, the country of his birth honored him by making the funeral a State function of the Yugoslav Government. Constantine Fotitich, Ambassador from Yugoslavia, represented his country.

The honorary pallbearers took their places. They were Dr. Ernest F. W. Alexanderson of the General Electric Company, inventor of the Alexanderson alternator; Professor Edwin H. Armstrong, inventor of a frequency modulator and other important radio devices; Dr. Harvey Rentschler, director of the research laboratories of the Westinghouse Company; Colonel

Henry Breckenridge; Gano Dunn, president of the J. G. White Engineering Corporation; Dr. Branko Cubrilovich, Yugoslav Minister of Agriculture and Supply; Consul General D. M. Stanoyevich of Yugoslavia; and Professor William H. Barton, curator of the Hayden Planetarium of the American Museum of Natural History, New York City. Heading this impressive list was Newbold Morris, president of the City Council of New York.

The Serbs and Croats seated themselves on opposite sides of the cathedral. Kenneth Swezey sat on the front row with Sava Kosanovich.

A hush settled over the crowd as the footsteps ceased. The stone walls of the building were in keeping with the discipline and dignity of the man whose body lay in the coffin. A flag of his adopted country was used instead of the customary pall.

Candles made flame halos from the tips of the tall tapers at the four corners of the casket. Their flickering light was caught and repeated in the gold ornamentation on the vestments of the priests. Tesla had been born by candlelight. He was leaving by candlelight. The ikon of the Most Holy Redeemer was set up at the choir steps. The officiants took their places for the Orthodox Service for the Dead.

Among the floral offerings which were set up around the casket was a wreath from King Peter II of Yugoslavia. The funereal scent of flowers mingled with the prayer-smoke of incense.

The casket was opened. Bishop Manning began the service in English, in the absence of Bishop Dionisije of the Serbian Orthodox Church.

The Very Reverend Dushan Shoukletovich continued the services in Serbian. Those to whom the foreign language was unintelligible sat quietly in their seats with their separate memories of this great man who had died. A myriad of thoughts and questions might have wandered into their minds:

"His long thumbed hands would be folded—and empty.
The shroud has no pockets and the dead go empty handed—
save for that which they have given away—
He had come to America with four cents!

He had given away millions with a single torn contract—
A million volts had passed through his body—
The power of Niagara was rushing through machinery that bore his name—
Where was the little device that could crumple the Empire State Building in thirteen minutes?
What if a tall tower had not fallen?
What could he have given the world, but they would not?
Where were the secrets of the death ray?
Stored in sixty musty trunks?
Engraved on the retinas of his closed eyes?
What of the night when he had been certain of signals from outer space?
Was he crazy?
How do you tell when a scientist is crazy?
What about his claim that he never made mistakes?
What of his wasted body, even now turning to dust?
What is death to an immortal?
What of his gigantic brain that had been short circuited by death?
Who would feed the pigeons in Bryant Park tonight—
and the next night—
and the next?
Was he down there now — running away again from something distasteful?
Was there a white dove with gray tips on her wings?
Gentle as a woman?
Were there unseen souls brooding over this mystic hour?
Was there a bearded one who once begged,
'Nikola, pray for me to that No-god of yours'?
Was the compassionate hand on his arm that of Madame Filopov?
Who was the form in the shadows who held so tightly to his purse strings?
Was the whispering voice that of Jules Verne saying,
'Anything one man can imagine, other men can make real.'
What of the Colorado night when Tesla, dressed in white tie and tails, had 'tamed a wild cat'—
and held the power of lightning in his hand?"

Bishop Manning pronounced the benediction over the body. Ambassador Fotitch led the procession past the coffin before it was closed.

The temperature was dropping and Kenneth Swezey and Sava Kosanovich buttoned their heavy coats tight and stepped into the car that was to follow the hearse to Ferncliff Cemetery at Ardsley-on-the-Hudson. A white building stood winter-stark among the bare trees. The high doors opened to admit the casket that was incredibly light. The body was placed in a receiving vault to await cremation.

The clouds cleared somewhat. There was a trace of snow in the cold air. The feeble sun had labored in vain to accomplish some afterglow of wintry glory for a genius whose sun had just set.

Once on the brink of discovery that was to power the world, Nikola Tesla had walked into a Budapest sunset, quoting, as though in a trance, the words of Goethe:

> The glow retreats, done is the day of toil;
> It yonder hastes, new fields of life exploring;
> Ah, that no wing can lift me from the soil,
> Upon its track to follow, follow soaring.
> A glorious dream! though now the glories fade.
> Alas! the wings that lift the mind no aid
> Of wings to lift the body can bequeath me.

Notes

1. Arthur J. Beckhard, *Electrical Genius, Nicola Tesla,* (New York, N.Y.: Julian Messner, Inc., 1959), 186.

2. Photostat of report from United States Department of Commerce, Weather Bureau.

3. Beckhard, *Electrical Genius,* 186.

4. Correspondence with Dr. Alan Fry.

5. Helen B. Walters, *Nikola Tesla, giant of electricity,* (New York, N.Y.: Thomas Y. Crowell, Co., 1961), 169.

6. Interview with Kenneth Swezey.

7. Hugo Gernsback, "The Cephalotaph" in *Forecast,* December 1959, 14-15.

8. Correspondence with Father E. N. "West, Canon of the Cathedral of St. John the Divine.

9. *New York Times,* January 13, 1943.

10. Ibid.

11. Correspondence with Father E. N. West.

12. *New York Times,* January 13, 1943.

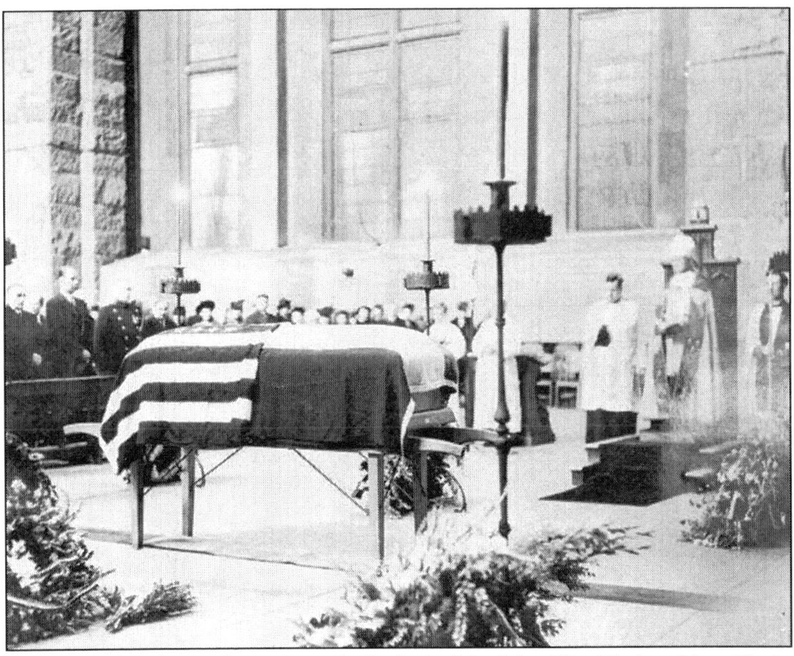

Nikola Tesla's funeral services, Cathedral of St. John The Divine, New York, January 12, 1943. *Top photograph From Kenneth M. Swezey Papers, Archives Center, National Museum of American History, Smithsonian Institution (AC0047-0000012). Bottom photograph courtesy of Fran Kiester.*

An exhibit at the Nikola Tesla Museum, Belgrade, Serbia, of a high-frequency oscillator and transformer built in 1955 from the technical descriptions of Tesla's Colorado Springs experiments. A Dickson D. Alley photograph of Tesla seated below his 1899 sparking oscillator is in the upper right. *From Kenneth M. Swezey Papers, Archives Center, National Museum of American History, Smithsonian Institution (AC0047-0000015).*

BIBLIOGRAPHY

Adams, William Simmonds, *Edison, His Life, His Work, His Genius.* Indianapolis, Ind.: Bobbs Merrill Co., 1934.

Bailey, Stephen, *L. L. Nunn, a Memoir.* Ithaca, N.Y.: Cayuga Press, 1933.

Beckhard, Arthur J., *Electrical Genius, Nicola Tesla.* New York, N.Y.: Julian Messner, 1959.

Bergaminia, David, *The Universe.* New York, N.Y.: Time, Inc., 1962.

Bradford, Gamaliel, *The Quick and the Dead.* Boston and New York, N.Y.: Houghton, Mifflin, Co., 1929.

Bulwer Lytton, *The Coming Race.* Edinburgh and London: Blackwood & Sons, 1871.

Colorado Springs Directory, 1900.

Dickson, W. K. L. and Antonia, *Life and Inventions of Thomas Alva Edison,* New York, N.Y.: Thomas Y. Crowell Pub. Co., copyrighted, *Cassiers Magazine,* 1894.

Eberhart, Perry, *A Guide to Colorado Ghost Towns and Mining Camps.* Denver, Colo.: Sage Books, 1959.

Hammond, J. W., *Charles Proteus Steinmetz.* New York, N.Y.: Century Company, 1924.

Hammond, John Winthrop, *Men and Volts, The Story of General Electric.* Philadelphia, Pa.: J. B. Lippincott, 1941.

Heathcote, Niels H. de V., *Nobel Prize Winners in Physics, 1901-1950.* Life of Science Library, No. 30. New York, N.Y.: Schuman Co., 1953.

Hix, J. Emile, *Can a Man Live Forever?* Chicago, Ill.: Western News Co., Publishing Agents, 1898.

Hunt, Inez and Draper, Wanetta, *To Colorado's Restless Ghosts.* Denver, Colo.: Sage Books, 1960.

Johnson, Robert Underwood, *Remembered Yesterdays.* Boston, Mass.: Little, Brown & Co., 1923.

Johnson, Robert Underwood, *Songs of Liberty* (including Paraphrases from Serbian, after translation by Nikola Tesla with prefatory note by him on Serbian Poetry). New York, N.Y.: Century Co., 1897.

Johnson, Robert Underwood, *Your Hall of Fame.*

Josephson, Matthew, *Thomas Alva Edison.* New York, N.Y.: McGraw Hill Co., 1959.

Leupp, Francis, *George Westinghouse.* Boston, Mass.: Little, Brown & Co., 1918.

Marconi, Degna, *My Father, Marconi.* New York, N.Y.: McGraw Hill, Co., 1962.

Martin, Thomas Commerford, *The Inventions, Researches and Writings of Nikola Tesla with Special Reference to His Work in Polyphase Currents and High Potential Lighting,* 1894, from *Electrical Engineer.* Reproduced by Lee Engineering Company, 1952.

Swezey, Kenneth, ed., *Nasa Posta.* Vol. Dedie au Jubilee de Nicolas Tesla, Beograd, Yugoslavia: 1932.

Nikola Tesla, 1846-1943, Lectures, Patents and Articles. Beograd, Yugoslavia: Nikola Tesla Museum, 1956.

O'Neill, John J., *Prodigal Genius, The Life of Nikola Tesla.* New York, N.Y.: Ives Washburn, Inc., 1944.

Ormes, Manley D. and Eleanor, *Book of Colorado Springs.* Colorado Springs, Colo.: Dentan Printing Co., 1933.

Pioneers of the San Juan Country. Vol. 4, published by Sarah Platt Decker Chapter, NS, D.A.R., Durango, Colo.: Big Mountain Press, 1961.

Schmeckebier, Laurence, *Ivan Mestrovic, Sculptor and Patriot.* Syracuse, N.Y.: Syracuse University Press, 1959.

Storm, Margaret, *Return of the Dove.* Margaret Storm Publication, Baltimore, Md.: Millenium Publication, 1959.

Tesla, Nikola, *Experiments with Alternate Currents of High Potential and High Frequency.* New York, N.Y.: McGraw Hill, Co., 1904.

Tribute to Nikola Tesla, Letters, Articles, Documents. Beograd, Yugoslavia: Nikola Tesla Museum, 1961.

Walters, Helen B., *Nikola Tesla, giant of electricity.* New York, N.Y.: Thomas Y. Crowell, Co., 1961.

Waters, Frank, *The Colorado.* New York, N.Y.: Rinehart & Co., 1946.

Wells, H. G., *The First Men on the Moon.* London, England: George Newnes, Ltd., 1901.

Who's Who in America. New York, N.Y.: McMiilan, Co., 1903—1905—1943.

Woodbury, David O., *Beloved Scientist.* New York, N.Y.: McGraw Hill Co., 1944.

Pamphlets

Cornell Branch of Telluride Association, Ithaca, N.Y.

Daugherty, Roland, *Nikola Tesla, the Father of Alternating Current Electricity (A Biographical Adventure).* Unpublished collection, 1961.

Dr. Nikola Tesla, 1856-1943, bibliography comp. and ed. Leland Anderson, Tesla Society, Minneapolis, Minn.: 1956.

Electrical West, Seventy-Fifth Anniversary Edition. New York, N.Y.: McGraw Hill, 1962.

Jackson, Merrick, *George Westinghouse,* Builder of Modern Industry Series, Good Reading Rack Service, Inc., New York, N.Y.: Mercer Publishing Co., 1958.

"Nikola Tesla Centennial Observance," The Tesla Society, Minneapolis, Minn.: October 1, 1956.

"Photo Bibliography of Nikola Tesla," compiled by Tesla Society, Minneapolis, Minnesota: May 1, 1959.

Telluride Chamber of Commerce, Pamphlet.

Letters

Correspondence between Nikola Tesla and
Mark Twain
George Westinghouse, Jr.
George Scherff
J. Pierpont Morgan
Robert Underwood Johnson
(microfilm from Library of Congress)

Correspondence with

Leland I. Anderson, Tesla Society.

Gerald L. Burger, United States Department of Commerce, Weather Bureau.

Charles Chamberlin, Office of Assistant Director, United States Information Agency.

Jack Craddock, Technical Information Officer, National Bureau of Standards.

Roland H. Daugherty.
Frank Davis, Henry Ford Museum.
Peter Day, editor of *The Living Church.*
Dr. Alan Fry, National Aeronautics and Space Administration.
Hugo Gernsback, editor of *Electrical Experimenter* and other periodicals.
Richard B. Gregg.
Dr. E. R. Hardy.
Louise F. Kampf.
Royal Lee, president of Lee Engineering Company.
Kenneth A. Lohf, Assistant Librarian, Special Collections, Columbia University.
David C. Mearns, Chief of Manuscript Division, Library of Congress.
Robert P. Multhauf, Head Curator, Department of Science and Technology, Smithsonian Institute.
R. V. McGahey, Manager of Technical Publicity, Westinghouse Electric Corporation.
Roy McVay.
Dr. E. Taylor Parks, Officer in Charge of Research Guidance and Review, Historical Research, U.S. Dept. of State.
Lloyd D. Prowell, Acting District Director, United States Department of Justice, Immigration and Naturalization.
John B. Reubens, Technical Information Officer for U.S. Department of Commerce, National Bureau of Standards.
Dr. Erik Rudberg, for The Royal Swedish Academy of Science.
Meta H. Schneider, Ferncliffe Cemetery Mausoleum Co., Inc.
P. C. Schools.
Agnes Wright Spring, former State Historian of Colorado.
Mae W. Stabler, Director of Information, *New York Herald Tribune.*
Kenneth Swezey, Tesla authority.
Monk Tyson.
Frank Waters.
Father Edward N. West, Canon Sacrist of the Cathedral of St. John the Divine.
A. Zambelli, Assistant Press Councelor, Embassy of the Federal People's Republic of Yugoslavia.

Magazines

American Mercury
Cassiers Magazine
Chatauqtan
Colliers
Colorado Magazine
Colorado Quarterly
Coronet
Craftsman
Current Biography
Current Literature
Current Opinion
Electrical Engineer
Electrical Engineering
Electrical Experimenter
Electrical Review
Electrical West
Electrical World
Electrical World and Engineer
Electrician
Empire Magazine
Engineering News
Facts
Family Weekly
Fate Magazine
Forecast 1960
General Electric Review
Life
Literary Digest
Look
Medical Tribune and Medical News
Medical Tribune
Mountain Sunshine
Nation
New Republic
Newsweek
North American Review
Outlook
Popular Science
Power
Public Opinion
Reader's Digest
Review (Yugoslav Monthly Magazine)
Review of Reviews
Rosicrucian Digest
Saturday Review of Literature
Science
Science News Letter
Scientific American
Technical World
Time
World Today
World's Work

Newspapers

Colorado Springs Evening Telegraph

Colorado Springs Free Press

Colorado Springs Gazette

Colorado Springs Gazette & Telegraph

Colorado Springs Weekly Gazette

Denver Post

Independent

London Times

New York Sun

New York Herald Tribune

New York Times

The Street Railway Gazette

The New York Sun

The Telluride Times

The World

Nikola Tesla Museum, Belgrade, Serbia. *Courtesy of Fran Kiester.*

Acknowledgments

Individuals:

Gene Bacon
Dorothy Bartlett
Arthur Baylis
Grace Berger
Michael Birchenall
Dr. Leo Bortree
S. A. Carol-Chick
Bruch Clark
Henry Clausen
Raymond Colwell
Perry Eberhart
Ruth Ellingson
Lorene Englert
Lt. Colonel George V. Fagan
Wesley Fernald
C. C. Ford
Ethel Ford
E. E. Foster
Helen Foster
Bruce Gunn
Joe Hardy
Robert Heinlein
Julie Hoefer
Dr. Hollis
Margaret Houston Aro
Glen Jackson
Helen Jackson
Paul Jones
J. A. Knight
Dr. Veljko Korac
Pearl La Fair
Virginia Labonte
Lora Light
Ellsworth Mason
Marjorie McGowan
David Mearns
Ernest Montgomery
Merle Montgomery
Howard Olson
Chris J. Oschman
Clara Perley
Frank Riddle
Rita Ridings
Theodore Roesch
Joan Shinew
Grant Sinton
Frank Stevens
E. Taylor Parks
Alice Van Diest
M. J. Vawser
Gertrude Walcher
Lentford Whitmore
Robert Wilfley
Mrs. Grant Wilson
Arthur Wyatt

Libraries:

Colorado Springs Public Library
Denver Public Library
Iowa State College Library
Library of Congress
McClelland Public Library, Pueblo, Colorado
Miami Public Library
New York Public Library
St. Louis Public Library
Tutt Library, Colorado College
U. S. Air Force Academy Library
University of California at Berkeley, Library
and their staffs
and other interlibrary loan services

An exhibit of personal items and photographs. Nikola Tesla Museum, Belgrade, Serbia. *From Kenneth M. Swezey Papers, Archives Center, National Museum of American History, Smithsonian Institution (AC0047-0000016).*

INDEX

Acheson, Edward Goodrich, 91
Adams, Edward Dean, 87
Alexanderson, Ernest F. W., 258
  on Tesla, 229
Alley, Dickson D., 7, 104, 113, 130, 264
Alta Vista Hotel (Colo. Springs), v, 106
  illust. of, 107
Alternating current, 41–55
  Columbian Exposition, 71–77
  dangers, 44–46, 60–61
  demonstrations, 51
  Edison &, 35, 44–46, 49
  electric chair &, 50
  first plant, 60–62
  Germany, 64–65
  mines &, 59
  motors, 27–28
  Niagara Falls &, 89–93
  phases, 33
  polyphase system, 43, 75
  Telluride &, 57–69
  Tesla's concept of, 23
Aluminum Co. of America, 91–90
American Institute of Electrical Engineers, 235
American Telephone Co. (Budapest), 26
Ames, (Colo.), 59–60
  power plant, 63–64
    rotor photo, 67
Anson, John (fictional character), 194
Anson, Vivian (fictional character), 194–195
Antlers Hotel (Colo. Springs):
  photo of, 4
Armstrong, Edwin H., 258
Astor, John Jacob, 105
  photo of, 106
Atomic energy, 183, 226, 243
Austin, Louis Winslow
  on Tesla, 228
Awards & Honors, Tesla's, 225–241
  listed, 237–238
Balloons, 113–114
Barton, William H., 259
Batchellor, Charles, 31–32, 33, 34
  executioner, 45–46
Bauzin (mayor), 33–34
Behrend, Bernard Arthur, 151, 177–178, 225
  death, 184
  on Edison, 231
  on Tesla, 230–231
Belgrade, (Serbia):
  map with, 8
Bell, John C.
  in fiction, 194, 196
Benson, A. L., 99–100, 101
Beograd, (Yugoslavia). *See* Belgrade, (Serbia)
Bernhardt, Sarah, 191
Biblical miracles & symbols, 192
Birds, Tesla &, 177, 219, 225
  dove, 197–198
Boat, torpedo:
  remote controlled, 97
Bortree, Leo, 127
Bragg, William Henry, 172
  on Tesla, 227–228
Bragg, William Lawrence, 172
Brancovic family, 14–16
Breckenridge, Henry, 259
Brisbane, Arthur
  photo of, 78
  Tesla &, 79–87

Brown, A. K., 41–42, 86
Brown, Harold P., 45, 47–48
Brown, M. L., 159
Budapest, (Hungary), 26
Buffalo, (N.Y.), 89, 90
  electrified, 91–90
Butler, Benjamin, 59

Campbell, James, 59
Can a Man Live Forever?
  by J. Emile Hix, 193–197
Capacitors:
  photo of, 134
Capital punishment:
  electicity &, 45–48
  electric chair, 50
Carr, Otis T., 198–199
Century Magazine, 206, 212, 213
Chalmers, Allis, 165
Chambers, Carl C.
  on Tesla, 240
Cholera, 17
  folk remedy, 17–18
Clemens, Samuel. *See* Twain, Mark
Cochran, Bourke, 46
Cohen, Louis, 229
Coils, 133
  photo of, 118, 130, 133, 135, 203
Colorado Springs Co., 105
Colorado Springs, (Colo.), 105–117
  coil, photo of, 203
  creditors, 158
  curiosity of reporters, 115, 128–129
  death ray &, 246
  experimental station photo, 7, 104, 113, 117, 130, 133, 134, 135, 167
  Independence Day, 120
  lab described, v, 109
  lab location, v, 4, 110, 117, 200
  photo of, 4, 55
  population, 4
  power outage, 136, 137, 139
  Tesla arrived, 105–106
  Tesla's experiments in, 5–7, 141
   wireless trans., photo of, 139.
   See also Tesla Experiment Station (Colo. Springs)
Colorado Springs Electric Co., 105, 136
  power plant, photo, 137, 139
  threatened a suit, 158
Columbian Exposition, 71–77
  illust., 70
  lighting, 71–77
  photo of, 75
The Coming Race
  by Edward Bulwer Lytton
   Vril power, 109–110
Compton, Arthur
  on Tesla, 228
Compulsions, Tesla's, 31, 190
Condensers. *See* Capacitors
Connecticut Yankee in King Arthur's Court
  by Mark Twain, 207
Cornell University:
  Telluride House, 63
Crawford, Marion
  photo of, 103
Crawford, William, 105
Cubrilovich, Branko, 259
Currency & stamps:
  commemorating Tesla, 174
Curtis, Leonard E., v, 74, 105, 108, 114
Custer, Elizabeth Bacon, 208
Custodian of Alien Property, 256, 257
Czito, Julius, 163
Czito, Kolman, 5, 132, 134–136

Death ray, 198, 243–253
 Colo. Springs &, 246
 illust. of news headlines, 242
De Forest, Lee, 132, 158
 on Tesla, 229–230
Delmonico's Restaurant (New York), 52, 79, 80
Denver & Gulf Railway, v, 107
Dewey, George, 132
Discipline, Tesla's, 15–16
Dodson, Johnson
 fictional character, 194
Dowda, Sandra, ii
Dozier, Joseph, 111, 117
Draganic (Tesla family name), 9
Drake, Frank, 127
Draper, Brad, ii
Draper, Paul, ii
Draper, Wanetta W., ii, iv, v
Duffner, Carl J., 158
Dunn, Gano, 259
 on Tesla, 230
Dvorak, Antonin, 207

Eaton, E. G.
 [is possibly Edwin J. Eaton], 127
Edison General Electric Co., 44
 Colorado &, 65
Edison Medal, 176
 Tesla &, 176–185
Edison, Thomas Alva, 22, 32, 36–38, 205, 206
 alternating current &, 44–46, 49
 animals electrocuted, 45–46
 capital punishment, 45–46
 Cylinder Phonograph, photo, 39
 death, 183
 direct current, 23
 disinterest in A.C., 35
 education, 35
 in fiction, 194
 humor, 36
 incand. lamp, 82
 Nobel Prize &, 169–170, 172–173
 patent on incand. lamp, 73–74
 photo of, 39, 46, 176
 Tesla &, 35–38
Education, Tesla's, 13, 16, 35
 European, 21–29
Einstein, Albert, 226, 243
Electricity:
 Tesla describes, 92
Electrocution:
 animals &, 45–46, 62–63
 capital punishment &, 45–48, 50
Electro magnetic motor:
 diagram of, 29
El Paso Club (Colo. Springs), 105, 108
Engineers, electrical:
 Deep Springs School, 63
 education, 62
 Telluride Association, 62–63
 Telluride Institute, 62

Fears & phobias, Tesla's, 11–12, 189–190, 190, 255
 revulsion to pearls, 192
Federal Bureau of Investigation, 257
Ferncliff Cemetery (Hartsdale, N.Y.), 261
Filipov, Luka, 208–209
 illust. of, 209
 Johnson's nickname, 209
Filipov, Madame. *See* Johnson, Katharine McMahon
Fire:
 reopen lab, 87
 Tesla's New York lab, 86–85
Fleming, Arthur P. M.
 on Tesla, 238–239

Flight:
  flying machines, 164–165
  Tesla's obssession with, 10, 11
Fluorescent light:
  photo of, 103. *See also* Phosphorescent light
Ford, John
  on Tesla, 240–241
Ford Motor Co., 199
Fotitich, Constantine, 258, 261
Fry, H. Alan, 256
Funeral, Tesla's, 257–261
  photo of, 201, 263

Gambling, Tesla &, 23–25
Gaulard, Lucien, 42
Gavrilovitch, Oscar, 258
General Electric Co., 44, 87
  Columbian Exposition, 72–73
  Niagara Falls &, 90, 91
Germany:
  spies, 179
  World War I &, 166
Gernsback, Hugo, 257
  on Tesla, 232
Gibbs, John, 42
Gilder, Richard Watson, 206, 213
Goethe
  recited by Tesla, 27
Gold King Mine:
  Nunn &, 59
  power &, 58
Gospich, (Lika, Croatia). *See* Gospic, (Lika, Croatia)
Gospic, (Lika, Croatia), 12
  cholera, 17
  fire-fighting equip., 13
  map with, 8
Governor Clinton Hotel (New York), 220
Gramme Dynamo, 22
Gratz, (Styria, Austria). *See* Graz, (Styria, Austria)
Graz, (Styria, Austria), 21
  map with, 8
  photo of, 24
Great Barrington, (Mass.):
  first A.C. lighted city, 42, 69
Gregg, Richard, 114–115

Hall of Fame of Great Americans, (New York):
  Johnson &, 218
Hall, Preston and Bobbitt, 105
Hammil, J. H., 246
Hardway, Glen, 248
Harper, Hugh D., 115–116
Health, Tesla's, 14–15, 26, 71, 86–87, 132, 164, 227
  accident, 255
  fainting spells, 220, 255
Higginson, Thomas Wentworth, 206
Higher Real Gymnasium (Karlovac, Carlstadt, Croatia):
  Tesla at, 14
Holden, Edward Singleton, 125–126
Holmgren, E. J.
  on Tesla, 240
Horseshoe Falls (Canada):
  photo of, 88
Hotel New Yorker (New York), 255
Hughes Aircraft, 251
Hunt, Inez, ii, iv, v
Hydroelectric power. *See* Niagara Falls

Immortality:
  in fiction, 195
Incandescent lamp:
  Columbian Exposition, 73–74
  Edison's, 82
  legal challenge, 74

Independence Day:
Colo. Springs celeb., 120
Cripple Creek celeb., 124
Interplanetary communication, 124–127, 148, 149, 193, 216, 227, 251-252
"In Tesla's Laboratory," (poem) by Robert Underwood Johnson, 221-222
Inventions & Patents, Tesla's, 85–86
A.C. motors, 27–28
alternating current, 23
"cold fire," 181
death ray, 243–253
discipline &, 16–15
disk turbine, 11
electro magnetic motor, diag., 29
failed to safeguard rights, 153
fluorescent light, photo, 103
flying machine turbine, 164–165
impact, v–vi
infringement, 66
mind-powered machines, 97–98
phosphorescent light, 207
radio, 215
remote control, 96–97
sold to Westinghouse, 42–43
speculations, 18, 83, 143, 145, 147–149, 152, 169, 170, 182-183, 193, 207, 244
television, 216–217
turbines, 163–167
warfare, 97
water turbines, 15–14
water wheel from disk, 11
wireless transmission of energy, photo, 55
world power system, 226

Jackson, Glen, 128
Jackson, Helen Hunt, 206
Jelenković, Vladimir, ii
Johnson, Katharine McMahon
death, 218
dying request, 218, 219
ill, 217
photo of, 217
Johnson, Robert Underwood, 85, 91, 128, 145, 147, 205–223
Ambassador to Italy, 216
death, 221
Hall of Fame, 218
health, 215
Ibsen Centenary, 220
"In Tesla's Laboratory," (poem), 221–222
letters to Tesla, 209–214
love of poetry, 208
money &, 213–219
photo of, 204, 216
Serbian poetry, 217–218
Journalism, yellow, 95, 187
Jovanovic Zmaj, Jovan, 208

Kelvin, Lord, 89
photo of, 90
Kemmler, William, 46
Kennelly, Arthur E., 45
Kentner, S. M.
on Tesla, 230
Kiester, Fran, ii
Kipling, Rudyard, 207–208
Knob Hill (Colo. Springs):
photo of, 110
Tesla Exp. Station &, 5, 158
Korn, Arthur
on Tesla, 230
Kosanovic, Sava N., 257, 259, 261
photo of, 254
Kuberski, Nina, ii

Laser:
  described, 248–251
Lenin, Vladimir Ilyich, 251
Lightning, 111, 121, 122, 256
  Colorado &, 119
  "Sky Fire," 244
Locksteadt, Charles, 72
Lowenstein, Fritz, 108, 114, 123
Lucan, John, 198

Maddocks, Chalmers E., 158
Madison Square Garden (New York):
  demonstration at, 97
Magnifying transmitter, 131–132
Maiman, Theodore H.
  laser, 248
Malaria:
  Tesla contracted, 14
  treatment, 16
Manning, William T. (bishop), 257, 259, 261
Marconi, Guglielmo, 99, 133
  communication with planets, 216
  law suit, 214
  photo of, 215
Marold, Richard, ii
Marriage, Tesla &, 191
Mars:
  communication with, 109, 111, 124–126, 148, 193
Martin, Thomas Commerford, 188, 205
  on Tesla, 238
Mascart, Eleuthere-Elie-Nicolas
  photo of, 90
Maser, 247–248
Matthews, Grindell H., 244–245
McAllister, Ward, 79
McClurg, Virginia, 109, 110
McDonald, Carlos F., 45
Mestrovic, Ivan
  bust of Tesla, 180–181, 234
Millikan, Robert
  on Tesla, 228
Mitchell, S. Weir
  fictional character, 194
Money, Tesla &, 25, 86, 128, 141–145
  borrowed, 142, 151, 159
  creditors, 153, 158–159
  digging ditches, 41
  first Tesla made in U.S., 35
  honorarium, 233–234
  Johnson &, 213–219
  Morgan &, 153–159
  never paid, 34, 37–38
  paid off past debts, 155, 255
  patents sold to Westinghouse, 42–43
  robbed, 34
  surrender of royalties, 50
  Wardenclyffe &, 151
Moore, D. McFarland
  on Tesla, 230
Morgan group, 87
Morgan, John Pierpont, 143–145, 151
  photo of, 143
  Tesla requesting funds, 153–159
Morgan, John Pierpont, Jr., 159
Morris, Newbold, 259
Moses, 192
Mysticism, 111, 201. *See also* Occult, Tesla &.

National Institute of Immigrant Welfare:
  scroll of honor given Tesla, 234
Natural Science Institute:
  fiction, 194–195, 196–197
New York City, (N.Y.):
  electified, 91

Niagara Falls and Conduit Co., 91
Niagara Falls (Canada), 153
  photo of, 88
Niagara Falls Hydraulic Power & Manufacturing Co., 88
Niagara Falls, (N.Y.), 6, 83
  A.C. power generation, 89-93
  General Electric &, 90-91
  Tesla's first knowledge, 15
  Westinghouse &, 90, 91
Nikola Tesla Co., 213
Nikola Tesla Day, 235
Nikola Tesla Museum (Belgrade, Serbia), ii, 234, 257
  photo of, 264, 270, 272
Nobel Prize (in physics):
  Edison &, 169-170, 172-173
  illust. of headlines, 168
  Tesla &, 169-175, 178, 214-215
Numbers, Tesla &
  divisible by three, 106
Nunn, Lucien Lucius, 57-58
  Gold King Mine &, 59
  photo of home, 56
Nunn, Paul N., 59, 62

Oborne, Ernest, 37
Occult, Tesla &, 3, 111, 149, 187-203
Olmstead Power Station, 63
O'Neill, John J., 197, 198, 200-201
  on Tesla, 239
Ophir, (Colo.), 59
Order of the White Eagle, 233
Otto, Calvin P.
  dedication, iv
  photo of, iv
Overton, Nellie, 158
Overton, Pearl, 158
Overton, Ruth, 158

Paderewski, Ignace Jan, 207
Paris, (France):
  Tesla on, 31
Parker, Guy, 114
Patents. *See* Inventions & Patents, Tesla's
Patterson, John
  on Tesla, 238
Peace:
  flying machine &, 147
Peck, Charles, 86
Peter II, King
  photo of, 254
Phosphorescent light, 207
  photo of lamp, 208
Pikes Peak (Colo.), 121
  firworks, 120, 124
Pikes Peak Ave. (Colo. Springs), photo of, 4, 110
Pikes Peak Library District, iii
Pittsburgh Reduction Co., 91-90
Poeschl, Jacob, 22-23
Pollen, Charles A., 158
Polytechnic School, Joanneum, (Graz), 21
  professors, Tesla's, 22
Populist Party:
  propaganda, 196
Portland Co., 65
Primozic, Franc, 236
Provo Canyon, (Utah), 63
  Olmstead Power Station, 63
Psychic phenomena, 84
Pueblo, (Colo.):
  in fiction, 194
Puskas, Tivadar, 31

Radar, 249
Radio, 215
Remote control, 96-97
Rentschler, Harvey, 258

Resonance, 99–100
  earth's, 100–101
  Houston Street lab, 101–102
Richards, Theodore W., 173
Rockwell, A. D., 45
Roentgen, Wilhelm, 84–85
  in fiction, 194
  photo of, 85
Roosevelt, Anna Eleanor
  condolences, 258
Rudolph, Katie, ii

San Miguel Consolidated Mining Co., 63–64
Scherff, George, 108, 156, 255–256
  injured, 152–153
Schley Rock, 121
Schley, Winfield Scott, 111, 112
Schoellkoff, Paul A.
  on Tesla, 239
Scott, Charles F., 90–91, 225
  on Tesla, 230, 239
Scribner's Monthly. See Century Magazine
Segvich, Stephen
  bust of Tesla, 235
Sellers, Coleman
  photo of, 90
Senses:
  Tesla's, 26, 119, 187
Severinghaus, W. L., 245
Shadowgraph. *See* X-rays
Shoukletovich, Dushan, 258, 259
Smiljan, (Lika, Croatia), 8, 9, 10
  photo of Tesla's home, 19
Smullin, Louis, 248
Societe Continentale Edison of France, 31
Sogg, Richard, 235
Soule, Gardner
  on Tesla, 240
South Side Machine & Metal Works:
  Columbian Exposition, 72
Space travel, 183
Spanish American War, 97, 111–112
  Tesla's inventions &, 98–99
S.S. City of Richmond (ship), 30, 34, 284
S.S. Oregon (ship), 35–36
Stanoyevich, D. M., 259
Stanton, Desire. *See* McClurg, Virginia
Stationary waves, 123
Statue of the Republic:
  photo of, 75
Steinmetz, Charles Proteus, 218
Stetson, Lyndle, 65
Stevens, Fred, 129
Stevens, Julie, 107, 112, 116
St. John Plumbing Co., 158
St. Regis Hotel (New York), 225
  suit against Tesla, 225
Supernatural. *See* Occult, Tesla &
Swezey, Kenneth M., 98, 177, 178, 184–185, 200, 225, 227, 256–257, 259, 261
  on Tesla, 231–232, 240
Szigety, Antal, 27, 33

Tatum, Edward, 45
Teleforce, 243–253
Telegeodynamics, 102
Telephone Exchange of Budapest, 31
Television, 216–217
Telluride Association, 62–63
Telluride, (Colo.):
  alternating current &, 57–69
  photo of Nunn home, 56
  power plant, 60
  Westinghouse &, 60

Telluride House:
Cornell &, 63
Telluride Institute, 62, 63
photo of, 69
Telluride Power Co., 64–63
Telluride Power Transmission Co., 64–63
Tesla, Angelina, 10
Tesla coil, 66, 116, 228
Tesla, Dane, 10
death, 12
Tesla, Djouka Mandich, 9–10
death, 71
Tesla Electric Co., 86
Tesla Experiment Station (Colo. Springs):
building described, 109, 116, 117
coil, photo of, 203
Colo. Springs, 5–7
deserted, 158
location, v, 4, 110, 117, 200
photo near, 4
photo of, 7, 104, 113, 117, 118, 130, 133, 134, 135, 167
sold, 158
wireless trans., photo of, 139
Tesla Institute (Belgrade, Serbia), 233
Tesla, Marika, 10
Tesla, Milka, 10
Tesla, Milutin, 9
apparent lack of appreciation, 21
aspirations for son, 14, 17
photo of Smiljan church, 19
Serbian Orthodox priest, 10
Tesla, Nikola, 95
accident, 255
alternating current, 23–22
arc light, 41
arrived in Colo. Springs, 105–106
awards & honors, 225–241
birds &, 177, 197–198, 225
birth, 9, 10
birthday, 75th, 225
birthday, 78th, 246
Brisbane &, 79–87
Budapest, 25–26
cholera contracted, 17–16
Columbian Exposition, 71–77
commem. currency & stamps, 174
compulsions, 31, 108, 190
cremation, 261
death, 200–203, 256
death mask, 257, photo of, 285
demonstrations, 51, 53, 71, 75–76
demonstrations, illust., 51
described, 79–80, 227, 285
developments foreseen, 6
diet, 14–15, 52
discipline, 15–16
ditch digger, 41
early childhood, 9–19
Edison &, 35–38
Edison's death, 183
education, 13, 16, 21–29, 35
electricity described, 92
estate, 256
experiments in Colo. Springs, 5–7
family, 9–19
father's aspirations, 14, 17
fears & phobias, 11–12, 116, 189–190, 190
in fiction, 194
fire-fighting equip. fixed, 13
funeral, 258–259
funeral, photo of, 201
gambling &, 23–25
greatest discovery, 122–123
health, 14–15, 26, 164, 227, 255
Higher Real Gymnasium, 14

Tesla, Nikola, (continued)
  Houston Street lab, 101–102
  illust. of, 51, 94
  letters to Johnson, 209–214
  light described, 82
  love of poetry, 208
  malaria contracted, 14
  manifesto, 140
  marriage &, 191
  ministry &, 12, 17
  money &, 25, 37–38, 141, 213–219
  movies &, 220–221
  mysticism, 111, 201
  Niagara Falls, 15
  Nobel Prize &, 169–175
  numbers div. by three, 106
  obsession with flight, 10, 11
  occult &, 3, 111, 149, 187–203
  passport app., photo of, 284
  philosophy &, 145–146
  photographic mind, 86
  photo of, 20, 30, 40, 81, 103, 130, 133, 162, 186, 224, 254
  photo of Smiljan home, 19
  Pittsburgh, 43
  polyphase system, 43, 75
  praise for others, 92–91
  precognition, 16
  rat-killer, 16
  read incessantly, 18, 22
  recluse, 182, 225
  records, 200
  religious beliefs, 137
  remote control, 96–97
  research resources, 3
  robbed, 34
  rotating magnetic field, 83
  royalties, 50
  scholarship, 21
  senses, 26, 119, 187, 199
  Serbian origins, 188
  shyness, 12
  social circles &, 52–53, 71
  spankings when child, 11
  Strassburg, 32–33, 33–34
  traveled to America, 34
  U.S. citizenship, 66, 218
  visions, 13–14, 23, 25, 27–28, 188–189
  war, opinion about, 112
  Westinghouse &, 42–43, 50, 141–142
  women &, 53, 84, 189, 190–192
Tesla (unit of magnetic flux density), 235
Thomas, Charles S., 109
Thomson, Elihu, 66
Thomson-Houston Co., 44, 65
  Columbian Exposition, 73
Thomson, William. *See* Kelvin, Lord
Time, 225
Tito, Josip Broz, 236
Townes, Charles Hard
  maser, 247–248, 251–253
Tributes, to Tesla. *See* Awards & Honors, Tesla's
Turbines, 163–167
Turrettini, Theodore
  photo of, 90
Twain, Mark, 207
  photo of, 208
  read by Tesla, 18

Union Printers Home (Colo. Springs):
  photo of, 104
Universal laws, 146
Unwin, William Cawthorne
  photo of, 90
Utah Power and Light Co., 64–63

Venus:
  communication with, 124, 148
  origin of Christ & Tesla, 197

Vibration. *See* Resonance
Virginius Mine, (Colo.):
  power &, 65
Visions, Tesla's, 13–14, 71, 188–189
Voltaire
  read by Tesla, 22

Waite, Davis Hanson, 196
Waldorf-Astoria Hotel (New York), 52
  photo of, 52
  Rose Room, 164
Wall, T. F., 245
Wardenclyffe Tower, (N.Y.), 150–152, 179
  costs, 151
  destruction, 179
  funding, 212–213
  halted, 159–160
  illust. of, 140
Warden, James S., 150
Waters, Frank
  on Tesla, 239–240
West, Edward, 258
Western Colorado Power Co., 64–63
Western Electric Co.:
  Columbian Exposition, 73
Westinghouse Electric and Manufacturing Co., 42, 50
  Columbian Exposition, 72–73
  incand. lamp, 73–74
  Niagara Falls &, 90, 91
Westinghouse, George, Jr., 42, 50
  buys Tesla patents, 42–43
  electricity for capital punish. &, 46–47
  photo of, 77
  safety of A.C., 48–49
  Telluride &, 60
White, J. G., 259
White, Stanford, 150, 179
Wilhelm, W. W.
  on Tesla, 240
Wireless:
  communication, 83, 84, 106, 152, 171
  transmission of energy, 181–182
    photo of, 55, 139, 162
Women:
  new sex order, 182
  Tesla &, 53, 84, 189, 190–192
World's Fair. *See* Columbian Exposition
World War I:
  mathematical equation, 165–166

X-rays, 84–85
  hazard, 153

Form for Naturalized Citizen.—[Edition of July, 1883.] No. 178. 100 W. Reid Gould, Law Blank Publisher and Stationer, 139 Nassau Street, cor. of Beekman, and 120 Broadway, N.Y.

No. 31959 Issued July 31/91

United States of America.

State of New York }
County of New York } ss:

I, Nikola Tesla a Naturalized and Loyal Citizen of the United States, do hereby apply to the Department of State at Washington for a passport for myself ~~born at ______ on the ______ day of ______ 18__, and~~

I do solemnly swear that I was born at Lika in Austria on or about the 15 day of June 1856; that I emigrated to the United States, sailing on board the S.S. City of Richmond from Havre on or about the 1 day of June 1884; that I resided 7 years, uninterruptedly, in the United States, from 1884 to date at New York City that I was naturalized as a citizen of the United States before the Court of Common Pleas at New York City on the 30th day of July 1891 as shown by the accompanying Certificate of Naturalization; that I am the identical person described in said Certificate; that I am domiciled in the United States, my permanent residence being at New York City in the State of New York where I follow the occupation of Civil Engineer; that I am about to go abroad temporarily; and that I intend to return to the United States in about six months with the purpose of residing and performing the duties of citizenship therein.

OATH OF ALLEGIANCE.

Further, I do solemnly swear that I will support and defend the Constitution of the United States against all enemies, foreign and domestic; that I will bear true faith and allegiance to the same; and that I take this obligation freely, without any mental reservation or purpose of evasion: So help me God.

Sworn to before me, this 30 *day of* July 1891

Nikola Tesla

Joseph H. Brown
Notary Public. Com of Deeds N.Y. Co

DESCRIPTION OF APPLICANT.

Age, 35 years.
Stature, 6 feet 1 1/2 inches, Eng.
Forehead, medium
Eyes, light brown
Nose, prominent
Mouth, medium with moustache
Chin, round
Hair, dark brown
Complexion, fair
Face, oval

IDENTIFICATION.

July 30th 1891

I Hereby Certify, that I know the above named Nikola Tesla personally, and know him to be the identical person referred to in the within described Certificate of Naturalization, and that the facts stated in his affidavit are true to the best of my knowledge and belief.

Richard Feist

Address of Witness, Rahway N.J.

Applicant desires passport sent to following address:

Nikola Tesla
Hotel Gerlach W. 27th Street
New York City

A death mask of Nikola Tesla was commissioned by Hugo Gernsback, who, as a memorial to commemorate the centennial of Tesla's birth, had it electroplated with copper in 1956. A second death mask and a gold-plated sphere containing Tesla's ashes are exhibited at the Nikola Tesla Museum, Belgrade, Serbia. *Courtesy of Larry Steckler, © 1956 by Radio-Electronics Magazine, New York.*

Nikola Tesla valued his U.S. citizenship papers more than any of the honors he received. On the very date of his naturalization, July 30, 1891, Nikola Tesla applied for a passport as new citizen of the United States. Among the details listed on the application (left), Tesla is described as having dark brown hair, light brown eyes, and a stature of 6 feet 1 ½ inches. Note that his birth date is stated to be on or about June 15, 1856—more than three weeks prior to the date commonly thought to be Tesla's birth date. *National Archives and Records Administration, NARA M1372. U.S. Passports, 1795 – 1905.*

would you have enjoyed
this book as much if it ~~we~~
had been fiction — the fact that it was
not fiction makes it all the more amazing
would you have ~~thought~~
all his inventions - alternating
current, the ~~empowering~~ of
recreating lightning in C.S. ~~purge~~
niagra falls, ~~impossible to~~
ridiculous fiction
believe if it had been real.

had it been fiction with all the
discoveries make believe
you might have said
that's not plausible
even for Science fiction